Accidental Exposure to Dioxins

Human Health Aspects

Academic Press Rapid Manuscript Reproduction

Proceedings of an International Forum on Human Health Aspects of Accidental Chemical Exposure to Dioxins—Strategy for Environmental Reclamation and Community Protection, October 4-7, 1981, Bethesda, Maryland

Organized by the International Academy of Environmental Safety (IAES) and the International Society of Ecotoxicology and Environmental Safety (SECOTOX) jointly with the Istituto Superiore di Sanità, Rome, Italy, and the Gesellschaft für Strahlen- und Umweltforschung MbH, Munich, Federal Republic of Germany

ECOTOXICOLOGY AND ENVIRONMENTAL QUALITY SERIES

Series Editors: *Frederick Coulston*
and
Freidhelm Korte

A complete list of titles in this series appears at the end of this volume.

ACCIDENTAL EXPOSURE TO DIOXINS

HUMAN HEALTH ASPECTS

Edited by

FREDERICK COULSTON

White Sands Research Center
Albany, New York

FRANCESCO POCCHIARI

Istituto Superiore di Sanità
Rome, Italy

1983

ACADEMIC PRESS
A Subsidiary of Harcourt Brace Jovanovich, Publishers

New York London
Paris San Diego San Francisco Saõ Paulo Sydney Tokyo Toronto

United Kingdom Edition published by
ACADEMIC PRESS, INC. (LONDON) LTD.
24/28 Oval Road, London NW1 7DX

ACADEMIC PRESS, INC.
111 Fifth Avenue, New York, New York 10003

Library of Congress Cataloging in Publication Data
Main entry under title:

Accidental exposure to dioxins.

(Ecotoxicology and environmental quality)
Proceedings of an International Forum on Human Health Aspects of Accidental Chemical Exposure to Dioxins–Strategy for Environmental Reclamation and Community Protection, Oct. 4–7, 1981, Bethesda, Md., organized by the International Academy of Environmental Safety and the International Society for Ecotoxicology and Environmental Safety, jointly with the Istituto superiore di sanita, Rome, and the Gesellschaft fur Strahlen– und Umweltforschung MbH, Munich.
Includes bibliographical references and index.
1. Tetrachlorodibenzodioxin–Toxicology–Congresses. 2. Tetrachlorodibenzodioxin–Environmental aspects–Congresses. I. Coulston, Frederick. II. Pocchiari, Francesco. III. International Forum on Human Health Aspects of Accidental Chemical Exposure to Dioxins–Strategy for Environmental Reclamation and Community Protection (1981 : Bethesda, Md.) IV. International Academy of Environmental Safety. V. International Society for Ecotoxicology and Environmental Safety. VI. Series.
[DNLM: 1. Dioxins–Toxicity. 2. Environmental health. 3. Environmental exposure. WA 240 A171]
RA1242.T44A25 1983 615.9'512 83-6032
ISBN 0–12–193160–9

PRINTED IN THE UNITED STATES OF AMERICA

83 84 85 86 9 8 7 6 5 4 3 2 1

CONTENTS

Participants ix
Preface xi

SECTION I

Welcoming Remarks 1

1 Environmental Impact of the Accidental Release of Tetrachlorodibenzo-*p*-Dioxin (TCDD) at Seveso (Italy) 5
F. Pocchiari, A. Di Domenico, V. Silano, and G. Zapponi
DISCUSSION 37

2 An Overview on the Health Effects of Halogenated Dioxins and Related Compounds—The Yusho and Taiwan Episodes 39
G. Reggiani

3 Reclamation of the TCDD-Contaminated Area at Seveso 69
Luigi Noe
DISCUSSION 75

SECTION II

Opening Remarks 79

4 The Metabolism of 2,3,7,8-Tetrachlorodibenzo-*p*-Dioxin in Mammalian Systems 81
J. R. Olson, T. A. Gasiewicz, L. E. Geiger, and R. A. Neal

DISCUSSION 101

5 Microbial Degradation of TCDD in a Model Ecosystem 105
F. Matsumura, John Quensen, and G. Tsushimoto

DISCUSSION 137

6 Oxidative Control of Chemical Pollutants by Ruthenium Tetroxide 139
D. C. Ayres

DISCUSSION 147

7 Methods of Photochemical Degradation of Halogenated Dioxins in View of Environmental Reclamation 149
Donald G. Crosby

DISCUSSION 159

8 Experts, Authorities, and Prophets 163
Etcyl H. Blair

SECTION III

Opening Remarks 169

9 Ecologic Chemistry of Dioxins and Related Compounds 171
F. Korte

DISCUSSION 183

10 Building Reclamation after Dioxin Contamination by PCB Fires 185
Nancy Kim

DISCUSSION 187

11 Mechanisms of Carcinogenesis Related to TCDD 191
Albert C. Kolbye, Jr.

DISCUSSION 197

12 Teratological Considerations on Dioxins 201
H. Tuchmann-Duplessis

DISCUSSION 211

SECTION IV

Opening Remarks 213

13 Health Impact of the Accidental Release of TCDD at Seveso 215
Paolo Bruzzi

DISCUSSION 227

14 Health-Related Problems to TCDD 229
Barclay M. Shepard

DISCUSSION 231

15 Problems in Agriculture with TCDD 233
Philip C. Kearney

DISCUSSION 243

16 Use of Epidemiology in the Regulation of Dioxins in the Food Supply 245
Frank Cordle

DISCUSSION 257

17 Possible Consequences of Sharing an Environment with Dioxins 259
Donald G. Barnes

DISCUSSION 267

SECTION V

Opening Remarks 269

18 Birth Defects in the TCDD Polluted Area of Seveso:
Results of a Four-Year Follow-Up 271
Paolo Bruzzi

Concluding Discussion *281*
Frederick Coulston, Chairman

Index *291*

PARTICIPANTS

Numbers in parentheses indicate the pages on which the authors' contributions begin.

D. C. Ayres (139), *Department of Chemistry, Westfield College, Haempstead, London, England*

Donald G. Barnes (259), *Office for Pesticides and Toxic Substances (TS-788), Environmental Protection Agency, Washington, D.C. 20460*

Etcyl H. Blair (163), *Health and Environmental Sciences, The Dow Chemical Company, Corporate R&D, Midland, Michigan 48640*

Paolo Bruzzi (215, 271), *Istituto di Oncologia, Università degli Studi, Viale Benedetto XV, 10, 16132 Genova, Italy*

Frank Cordle (245), *Epidemiology Unit, Bureau of Foods, Food and Drug Administration, Washington, D.C. 20204*

Frederick Coulston (281), *White Sands Research Center, Albany, New York 12208*

Donald G. Crosby (149), *Department of Environmental Toxicology, University of California, Davis, California 95616*

A. Di Domenico (5), *Istituto Superiore di Sanità, Viale Regina Elena 299, Roma (00161), Italy*

John P. Frawley, *Hercules, Inc., 910 Market Street, Wilmington, Delaware 19899*

Peter B. Hutt, *Covington and Burling, 1201 Pennsylvania Avenue NW, Washington, D.C. 20004*

T. A. Gasiewicz (81), *Environmental Health Sciences Center, University of Rochester School of Medicine, Rochester, New York 14642*

L. E. Geiger (81), *Center in Toxicology, Vanderbilt University, Nashville, Tennessee 37232*

Philip C. Kearney (233), *Pesticide Degradation Laboratory, Agricultural Research Service, U.S. Department of Agriculture, Beltsville, Maryland 20705*

Nancy Kim (185), *Bureau of Toxic Substances Management, New York State Department of Health, Albany, New York 12237*

Albert C. Kolbye, Jr. (191), *Bureau for Toxicology, Bureau of Foods, Food and Drug Administration, Washington, D.C. 20204*

F. Korte (171), *Institut für Okologische Chemie der Gesellschaft für Strahlen- und Umweltforschung mbH München, Ingolstädter Landstrasse 1, D-8042 Neuherberg, Federal Republic of Germany*

F. Matsumura (105), *Pesticide Research Center, Michigan State University, East Lansing, Michigan 48824*

R. A. Neal (81), *Chemical Industry Institute of Toxicology, Research Triangle Park, North Carolina 27709*

Luigi Noe (69), *Ufficio Speciale per Seveso, Via S. Carlo 4, 20030 Seveso, Milano, Italy*

J. R. Olson (81), *Department of Pharmacology and Therapeutics, School of Medicine, State University of New York, Buffalo, New York 14214*

F. Pocchiari (5), *Istituto Superiore di Sanità, Viale Regina Elena 299, Roma (00161), Italy*

John Quensen (105), *Pesticide Research Center, Michigan State University, East Lansing, Michigan 48824*

G. Reggiani (39), *Medical Research Board, F. Hoffmann-La Roche & Co. Ltd., Grenzacherstrasse 124, CH-4002 Basle, Switzerland*

Barclay M. Shepard (229), *Environmental Medicine, Veterans Administration Central Office (102), Washington, D.C. 20420*

V. Silano (5), *Istituto Superiore di Sanità, Viale Regina Elena 299, Roma (00161), Italy*

G. Tsushimoto (105), *Pesticide Research Center, Michigan State University, East Lansing, Michigan 48824*

H. Tuchmann-Duplessis (201), *L'Academie Nationale de Medecine, Paris, France*

G. Zapponi (5), *Istituto Superiore di Sanità, Viale Regina Elena 299, Roma (00161), Italy*

PREFACE

On July 10, 1976, there was, in a city near Milan, Italy, an uncontrollable exothermic reaction in a factory, which released a cloud of steam and vapor into the atmosphere. For a period of several days, this accident was not considered serious until it was observed that birds, vegetation, and courtyard animals began to exhibit signs and symptoms of toxicity. Indeed, some of the people in the community exposed at this time began to complain about lesions on their skin. Nine days later, tetrachlorobenzo-P-dioxin (TCDD) was analyzed in various samples collected near the factory.

The purpose of this volume is to document the nature of the exposure to dioxin of plants, animals, and man and to discuss in great detail what was done to alleviate the situation and, indeed, what can be learned from this experience and how to manage any other accident of this kind in the future. The title of the conference that provided the material for this book was "Human Health Aspects to Accidental Chemical Exposure of Dioxins—Strategy for Environmental Reclamation and Community Protection." The meeting was organized and sponsored by the International Academy of Environmental Safety (IAES) and the International Society of Ecotoxicology and Environmental Safety (SECOTOX) jointly with the Istituto Superiore di Sanita, Rome, Italy, and the Gesellschaft fur Strahlen- und Umwelt Forschung mbH, Munich, Federal Republic of Germany. The chairmen were Dr. Frederick Coulston and Dr. Francesco Pocchiari.

Dioxin itself has never been used by man as a commercial product in our environment. Clearly, it is a contaminant in the production of useful chemicals and it can occur also as a result of combustion of organic material such as coal, oil, and wood. Because it is ubiquitous in our environment, the U.S. Food and Drug Administration has set an allowable value of 1 nanogram of dioxin per kilogram of body weight in food consumed, for example, in fish.

What happened in Seveso from a chemical point of view is not unique; there have been other mishaps in factories where emissions of dioxin have been released into the atmosphere. What is unique in the Seveso accident is that the emissions, instead of staying inside the factory, were spread outside over the landscape surrounding the factory. Indeed, at Seveso, a cloud of emissions followed the wind pattern in a classic way. Two major previous accidents have occurred: one in Germany and one in the

United States of America. Both these episodes have been studied very carefully from an epidemiologic point of view over a period of more than 30 years. As a result of these studies, much is known about the effects of a single exposure to humans resulting from an accidental emission of dioxin. In no case has there been any reported increase in the incidence of any major category of disease, including cancer, except for the presence of chloracne. Indeed, there is little or no evidence that any serious long-term effect was observed in any of the people exposed to levels of dioxin high enough to cause chloracne.

Questions as to the effect of dioxin on the reproductive process are discussed at some length in the report. There was much fear that pregnant women would suffer from an exposure to dioxin. The question of teratological effects and even miscarriages and stillbirths were obviously points of great concern. The presentations and discussions in this volume should help very much to answer this question in the negative. Whether the exposure to dioxin in the Seveso district will result in cancer many years from now, of course, could not be addressed at the meeting, but the experience with at least the two major emissions accidents mentioned above, in Germany and in the United States, goes a long way to giving scientific information regarding this question, namely, that there has been no increase in any form of cancer so far over a period of 20 to 30 years following exposure.

The environmental impact of this accident, the fate of dioxin in the environment, the methods used, the hygienic and protective methods employed, are discussed in great detail in this book.

The presentations by Dr. Francesco Pocchiari and his colleagues from Italy represent the official presentation of their data. They are to be highly complimented for their action and diligence in helping to solve the problem of dioxin in the Seveso area. Many international and national experts expressed the opnion that a disaster would occur, not only amongst the pregnant women, but amongst the entire population. However, the Italians set about to ascertain the nature and extent of the exposure and quietly went about correcting the situation. Indeed, as their report clearly states, no harmful effects were observed except for chloracne and perhaps the death of some small animals and plants. They are to be complimented for turning away from the ''doomsday'' aspects of what was supposed to happen, to the reality of what indeed did happen, namely, that the situation could be controlled by good and adequate methods of containment and cleanup.

The lesson that can be learned from the Seveso dioxin experience and, in retrospect, from the major German and United States accidents, is that the human species is much less sensitive to the dioxins than are rodents or other small animals. There is no obvious or serious long-term consequence resulting from an exposure to dioxin that was sufficient to cause chloracne in the human population. Above all, it clearly demonstrated that pregnant women were not unduly sensitive to the toxicity of the dioxin and that the question of cancer has apparently been resolved by virtue of the previous accidents where a 30-year followup has not revealed any increase in the cancer rate. In fact, it can be stated that there was a lower mortality rate in general amongst the people exposed in the 30-year-old accidents. There are 22 isomers of dioxin and many of these exist in our environment. There is no doubt that the dioxins are undesirable and everyone agrees that the dioxins are harmful chemicals. However, they are present in extremely low amounts in some of the chemicals and drugs that we use, and the question is what can be done to eliminate the dioxins from our society and environment as much as possible.

The meeting concerned itself primarily with the accidental exposure of plants, animals, and man to dioxins and, indeed, what can be done to clean up a community after such an exposure. The human health aspects of exposure to dioxins were considered and all of the major points were discussed that could be revealed by a study of the water, the air, the food and, indeed, the very earth that was contaminated by the dioxins. The only clear-cut indication of toxicity, in the more than 100 children (8 to 14 years of age) exposed to dioxin, was chloracne. No chemical, except the dioxins, could have caused this condition in the children and in some of the adults. It is known from human experiments that a dermal dose of 7500 μg can produce severe chloracne in man, but that a lower dose at the range of 1 μg does not. Unfortunately, chloracne persists for a long time and the nature of this lesion is not understood.

In a four-year post-exposure study on birth defects, it was clearly demonstrated that there was no real teratological effect, nor was there any apparent change in the birth rate. Whether the dioxins act as initiators or potentiators of cancers has to remain an open question until further studies are done. However, as pointed out previously, there has been 30 years' experience with the exposure of several hundred people in factories with no increase in cancer.

The decontamination procedures developed by Dr. Pocchiari and his colleagues are worthy of note and indicate for the future that it is possible to go into an area of contamination, such as occurred in Seveso, and clean it up. It appears that some animals, such as rabbits, are very sensitive to the dioxins and these can be used as an indicator for contamination. The methods of decontamination of buildings and the soil developed by the Italian group is revealed as a protocol for future control of such accidents. Indeed, these methods can be used for the general decontamination of the soil by waste materials containing some of the dioxin, PCB, and other chemicals. The emergency plan that evolved in Italy can indeed be useful in dealing with other situations worldwide and should be followed, not only by nations, but by industry and communities in which chemical plants are located, which may result in contamination of the environment. It is now recognized that there is a need for prompt action, not to wait, as in the case of the Seveso episode, for 10 days to pass before the extent of the tragedy was known.

At first it appeared that the Seveso accident was to become a disaster. In hindsight now, as pointed out by Dr. Pocchiari in his closing statement, it certainly was not a toxicologic disaster, but what it became was a public disaster, specifically in terms of buildings, agriculture, etc., and even the behavior of people, the press, and the media. The toxic effect on humans was limited to chloracne, but the dioxins were apparently toxic to animals and plants. Five years later the situation appears to be under complete control. An emergency plan in force at the time of the Seveso accident would have saved valuable time and contributed to the understanding of the extent of the tragedy and its management.

The experience and lessons learned from the emission of dioxin at Seveso have been carefully documented by Dr. Pocchiari and his colleagues. The lessons learned should contribute very much to our understanding of other similar situations, such as the presence of small amounts of dioxins in 2,4,5-T, Agent Orange, and even the Times Beach land contamination. Clearly, more scientific information is needed before the final story can be told about the dangers of dioxins in our environment.

WELCOMING REMARKS

DR. COULSTON: The purpose of this open meeting is to discuss in a detailed and comprehensive fashion, with the experts around the table, the problems concerning the dioxins related to the Seveso episode. This is really the first time the incident has been discussed in detail.

We are very fortunate in that this is an official meeting of the International Academy of Environmental Safety and The Society of Ecotoxicology in cooperation with the Istituto Superiore di Sanità of Rome; and it is a semi-official meeting of the Italian group represented around this table. It is urgent and timely that we listen to the true story, as far as it can be told today, to what actually happened at the time of the Seveso accident that released the gases that contained the dioxins. The key point of this meeting is absolutely free and open discussion.

PROFESSOR KORTE: I wish to speak on behalf of my colleagues from the Ministry for Research and Technology in the Federal Republic of Germany, for the Gesellschaft für Strahlen, und Umweltforschung, from the International Academy of Environmental Safety and from our little, but very active, Forum für Wissenschaft, Wirtschaft, und Politik, in Bonn. There are many organizations in Germany interested in these kinds of problems. From 15 to 20 percent of our income comes from chemistry, and we must be careful to understand chemistry as best we can. I think this meeting is very important and provides a good example of a model study. We should be able to assess chemicals, both natural and synthetic, in terms of their hazards. But no one knows how to do it. Therefore, the only way we can go forward

ISBN 0-12-193160-9

step by step is by using examples. Dioxins are excellent in this respect because they are not produced for anything. They are just present as accidental contaminants of chemical reactions that sometimes enter into the environment. These model studies, then, can teach us how to handle other chemical problems in the future. It is a very great task.

DR. COULSTON: Professor Pocchiari is the co-chairman of this conference. He heads an institute in Rome that is a combination of what we would call in the United States, the National Institutes of Health and The Center for Disease Control, and he would have responsibility for some of the regulatory activities of our EPA and FDA. In other words, he represents an institute whose aim is to protect the Italian people from the harmful effects of drugs, to establish safety standards in the use of chemicals and drugs, and to determine the efficacy of drugs and food additives and their safe use.

PROFESSOR POCCHIARI: As you know, after the Seveso accident many meetings have taken place. In the last 5 years these problems have been discussed in different parts of the world, including Italy, the United States, Sweden, Germany, and other countries. Today, we have a new opportunity to discuss the matters of rehabilitation. People are now interested in health as related to rehabilitation and community protection. This is very important because now we have an international program for chemical safety, the IPCS, organized 2 years ago by WHO. There is now a regional office in Copenhagen which was created to meet emergency situations. It has established guidelines for contingency planning for responses to chemical accidents involving potentially toxic substances. These plans will be published in the next few months. In the meantime, we are trying to develop plans and guidelines for the rehabilitation of areas already affected by chemical accidents. These plans will be very important to governments having to meet these emergencies. It is important to prepare guidelines for the future. Working in collaboration, we of the Istituto Superiore di Sanità in Rome and Professor Philip Jones of the Institute of Environmental Studies in Toronto have established some of these guidelines.

Therefore, it is important that at this meeting we discuss the problems of rehabilitation that we consider at this moment to be among our most important tasks.

DR. COULSTON: This meeting should end with a full understanding of the implications of the Seveso accident, and what can be done about it. Perhaps Dr. Pocchiari and his staff can tell us more about the plans of the WHO and EEC groups. Then, perhaps, someone from our own regulatory agencies can tell us if they have made similar plans. We have around this table representatives of almost every agency that one can think of as involved in issues of this kind. For this reason, we should be able to discuss what can be done both retrospectively and prospectively for an occasion such as Seveso.

There have been many accidents in the past involving similar instances such as Seveso. There was the one in Nitro, West Virginia, studied very carefully by Ray Suskind. There was one in Arkansas which was examined carefully by Dr. Irving Selikoff; the results are not yet published. There have been other chemical accidents around the world: accidents with vinyl chloride, accidental exposures to PCB's, PBB's, and you name it--it has all happened. There is one pervasive characteristic though: most of these are one-time accidents--they occurred and they were over. We hope to prevent other accidents from occurring. In many cases the chemical was withdrawn and the incident can never happen again -- like PBB. In the case of dioxins, on the other hand, not only does man involve himself with it as he makes it as a contaminant while he produces other products, but he does not make it for use, since it occurs as a side product of manufacturing processes. It occurs, one might say, in nature. It comes out of smoke stacks as materials such as coal, oil, and similar organic products are burned. It seems to be present in the environment in an ubiquitous way not necessarily related to anything that mad did *per se*. This distinguishes the dioxins as a problem. This issue will be discussed later by Dr. Barnes. How do you live in an environment that has dioxins in it and what do you do about it? It is in the food, in fish, air and it is going to continue to remain there for some time to come. This is one of the issues of this meeting that we want to address.

Now we will have the keynote address by Dr. Pocchiari.

CHAPTER 1

ENVIRONMENTAL IMPACT OF THE ACCIDENTAL RELEASE OF TETRACHLORODIBENZO-*p*-DIOXIN (TCDD) AT SEVESO (ITALY)

F. Pocchiari, A. Di Domenico, V. Silano, and G. Zapponi

Istituto Superiore di Sanità
Rome, Italy

I. INTRODUCTION

On July 10th, 1976, an uncontrollable exothermic process began somewhere in the reaction bulk during the synthesis of trichlorophenol at the Givaudan-La Roche ICMESA plant at Meda, 30 km north of Milan. A toxic cloud was thereby released into the atmosphere. About 1800 ha of a densely populated area, called the Brianza of Seveso, were contaminated by this toxic cloud (Fig. 1). Within a few days of the accident, vegetation, birds, and courtyard animals near the ICMESA plant were seriously affected. At the same time, dermal lesions began to appear among the inhabitants of the area. Only 9 days after the accident, it was assessed that tetrachlorodibenzo-*p*-dioxin (TCDD) was present in various types of samples collected near the ICMESA plant. As a first step, on July 26, 1976, Italian Authorities evacuated 179 people from a 15-ha area immediately southeast of the plant. A few days later, further sampling of soil and vegetation indicated the presence of TCDD, which

ISBN 0-12-193160-9

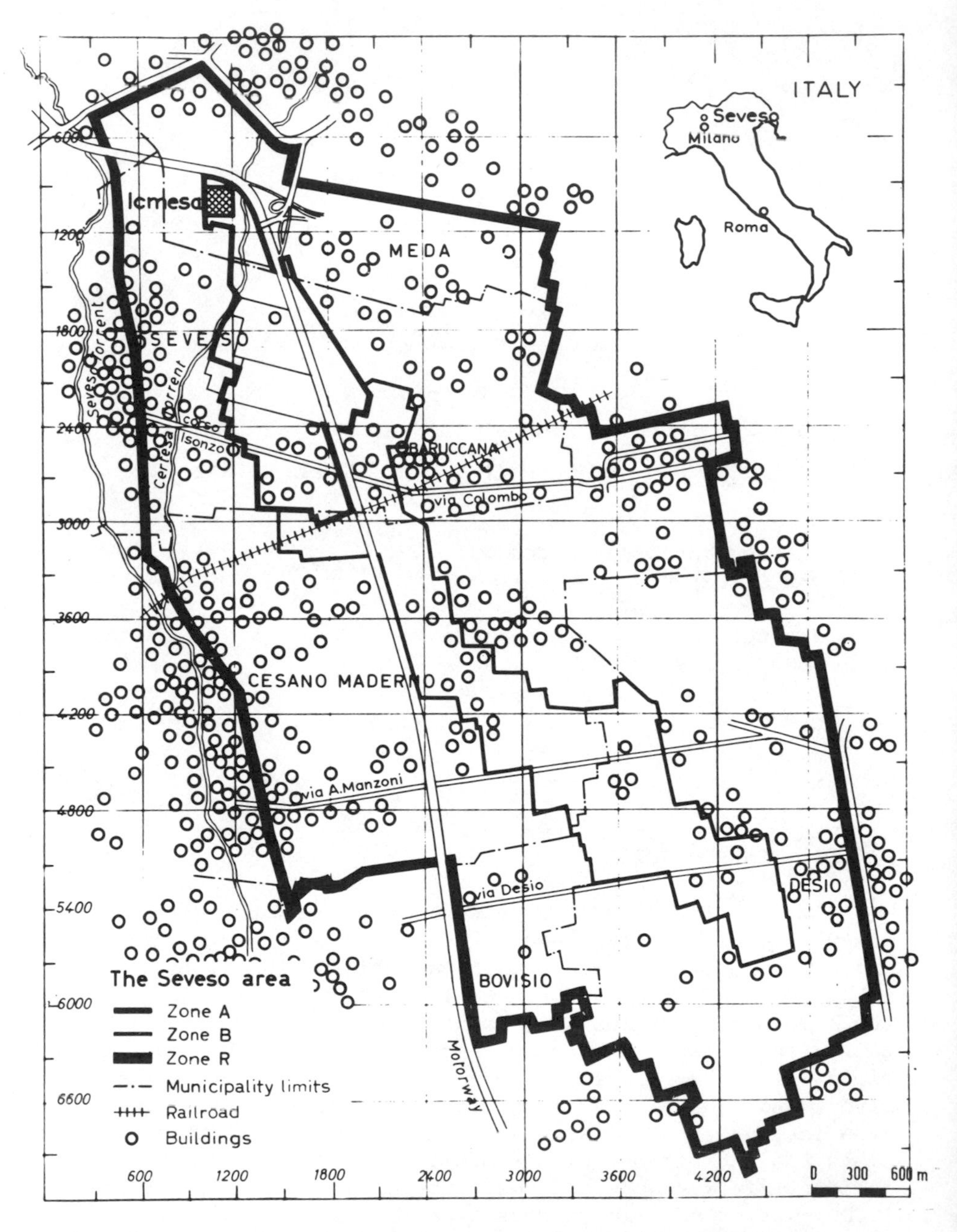

Fig. 1. Zones A, B and R at their maximum extension, showing major built-up areas (O) and surrounding farm lands. The ICMESA plant appears within Meda municipality boundaries near the Meda-Seveso borderline. Most of Zone A lies within the

prompted Italian Authorities to evacuate all the inhabitants (733 people) in a wide area, coded Zone A (approximately 110 ha)(Fig. 1). Zone B (270 ha), the natural extension of Zone A along the main diffusion pathway of the cloud, exhibited lower dioxin contents. Both Zones A and B were enclosed by a larger territory, coded Zone R (1430 ha). Zones B and R were subjected to a number of hygenic measures.

This chapter appraises the environmental impact of this accident five years following this event. The behavior and fate of 2,3,7,8-TCDD released into the Seveso environment is also discussed.

II. EXPERIMENTAL PROCEDURES

TCDD levels in soil, ground and surface waters, sediments, and airborne dusts were determined as reported by di Domenico et al. (1980a). TCDD levels were determined in vegetation as reported by Cavallaro et al. (1980), in plant seeds as described by Cavallaro et al. (in press), and in animal tissues and earthworms as reported by Fanelli et al. (1980a). TCDD levels in milk were given by Fanelli et al. (1980b). In most cases, TCDD levels were analyzed using GS-MS techniques.

III. RESULTS

A. TIME DEPENDENCE OF TCDD LEVELS IN THE SOIL OF ZONE A

The surface distribution of TCDD over sectors of Zone A, detected at different times following the accident, is shown in Figs. 2 and 3. Figure 2a shows the first contamination picture of the soil surface of Zone A. The clear portion of the map indicates the area where, due to emergency conditions, TCDD monitoring was performed mainly on vegetation

---

(Continuation Fig. 1)

Seveso municipal boundaries, while Zone B comes within the Cesano Maderno and Desio municipality boundaries. Torrent Certesa is an affluent of Torrent Seveso which, in turn, ends up in River Lambro, 30 km south, in Milan. River Lambro flows through Milan and is a tributary of River Po. Both rivers are south of the Zone R southern tip and, hence, are not shown. The reference grid is oriented north-south.

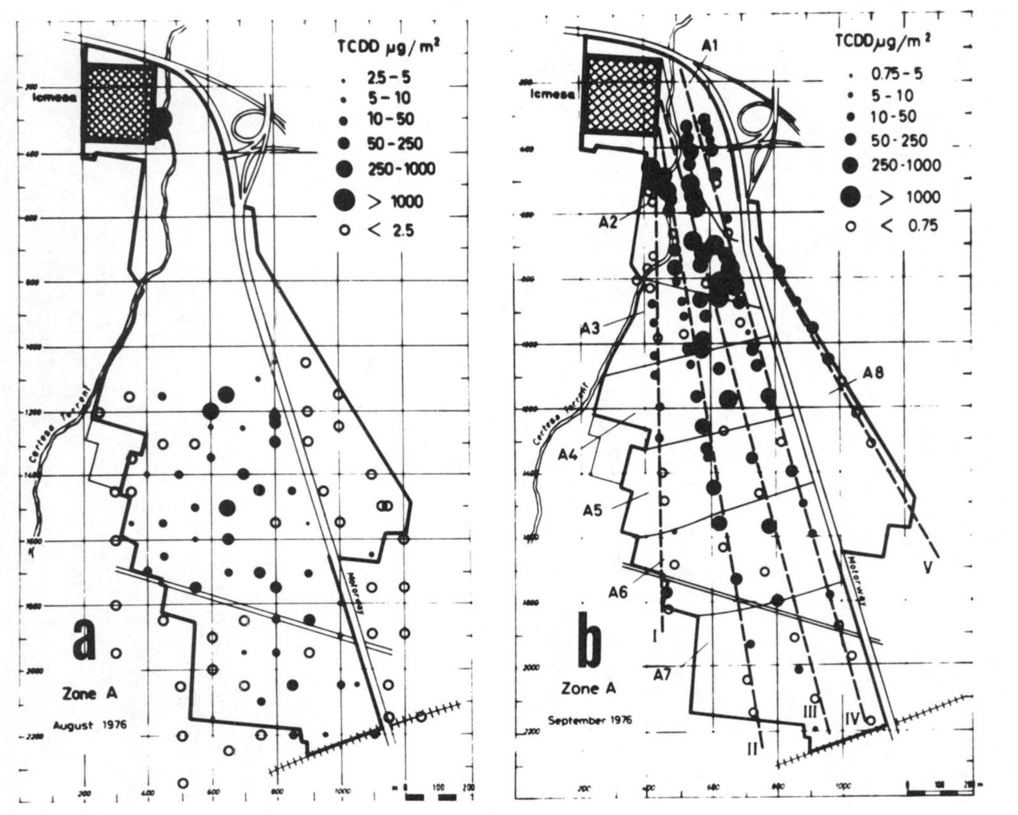

Fig. 2. Evolution in time of Zone A TCDD levels (a) The upper part (clear area) of the August, 1976 map was determined based on TCDD findings of various environmental samples collected before July 26, 1976. The boundaries of the lower part of Zone A (spotted area) were established on the basis of numerous top soil specimens sampled on July 26 and analyzed immediately afterward. The north-south square grid shown is intended to facilitate the comparison of TCDD levels in this map with those reported in (b), (c), and (d), but does not identify with original coordinates. Mapped site locations are accurate to ±50 m. Early analytical findings were expressed as μg/100 g of sample (Adamoli et al., 1978). They were here converted to μg/m² to make data values consistent with those of other maps. Data are accurate to approximately ±2.5 times the

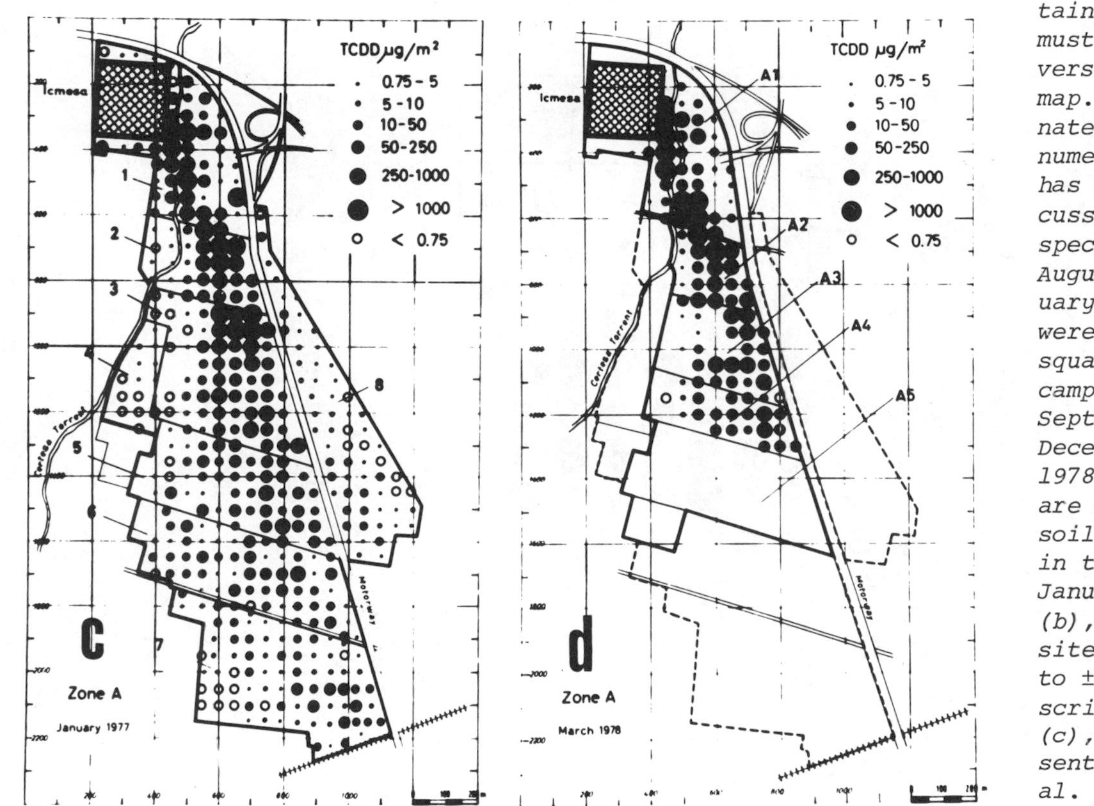

µg/m² numerical value obtained (not reported) which must be allowed for such conversion. *(b)* September 1976 map. Original polar coordinates are labeled with Roman numerals; the square grid has been superimposed as discussed above in *(a)*. Soil specimens were collected on August 11-13, 1976. *(c)* January 1977 map. Sampling sites were established on a 50-m square grid. The sampling campaign started in late September and lasted through December, 1976. *(d)* March 1978 map. Sampling sites are as in case *(c)*. Most soil specimens were sampled in the December 15, 1977-January 5, 1978 period. For *(b)*, *(c)*, and *(d)* mapped site locations are accurate to ±25 m. An extensive description of the maps *(b)*, *(c)*, and *(d)* has been presented by di Domenico et al. (1980b).

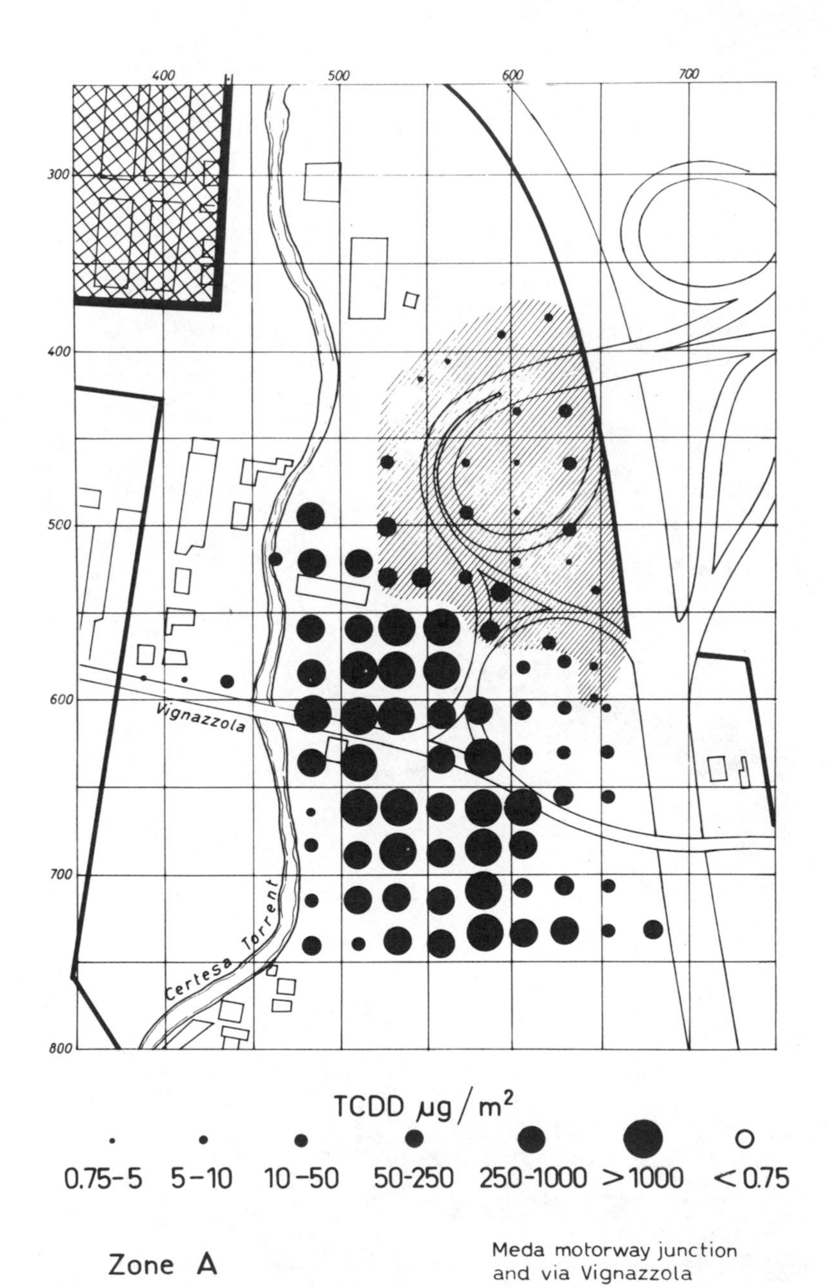
400
500
600
700
300
400
500
600
700
800
Vignazzola
Certesa Torrent
TCDD μg/m²
0.75–5
5–10
10–50
50-250
250-1000
>1000
<0.75
Zone A
Meda motorway junction
and via Vignazzola

Figure 3

samples that were clearly affected by the chemical action of the toxic cloud (see Section III,E). This map, together with other preliminary analytical findings and information available on the toxic and pathological events and on the airstream patterns at the time of the blow out, allowed the evacuation of the area at the end of July, 1976. Soil surface contamination of Zone A was assayed in more detail the second week of August, 1976 (Fig. 2b). Higher TCDD levels were present in the part of Zone A closer to the ICMESA plant. TCDD levels decreased with increasing distance from the ICMESA plant and/or east or west from the south-southeast direction (reference line III, Fig. 2b). This suggests that most of the chemical cloud was wind-driven along a definite direction and that, moving away from the main diffusion pathway and/or from the ICMESA plant, a gradual dispersion process occurred within the chemical cloud (di Domenico et al., 1980b).

Subsequent mappings of Zone A were completed by January, 1977 and March, 1978 (Fig. 2c and 2d). The 1977 map represents the most complete and systematic one available of Zone A (Fig. 2c). The 1978 map (Fig. 2d) refers solely to subzones A1-A4, since subzones A6-A8 had been detoxicated prior to its preparation and subzone A5 was being utilized as a deposit. This map was drawn mainly to collect additional information on the persistence of TCDD in the soil of this Zone. Last, Fig. 3 shows TCDD levels in the soil surface of the other sectors of Zone A that were recently reclaimed.

TCDD levels at 44 corresponding soil sites in Zone A during the three surveys carried out at 1, 5, and 17 months after the ICMESA accident provided statistically significant ($p<0.01$) evidence, which indicated that the geometric mean of TCDD in the soil of Zone A, left unworked since the accident, diminished sharply in the first 6 months after the accident. Following this period, no further significant decrease in TCDD levels were detected. Available data are consistent with two mathematical models (Fig. 4). By extrapolating from one of these models, TCDD levels present in soil immediately after the accident were estimated to have been from 5 to 11 times

---

Fig. 3. TCDD levels in two sectors of Zone A: the Meda motorway junction (shaded area) and Via Vignazzola. Soil samplings were carried out on October 13, 1978, for the former sector, and from April 10 to June 3, 1980, for the latter. Sites were established on the basis of the reference grid previously utilized (Figs. 2c,d), but their frequency was doubled.

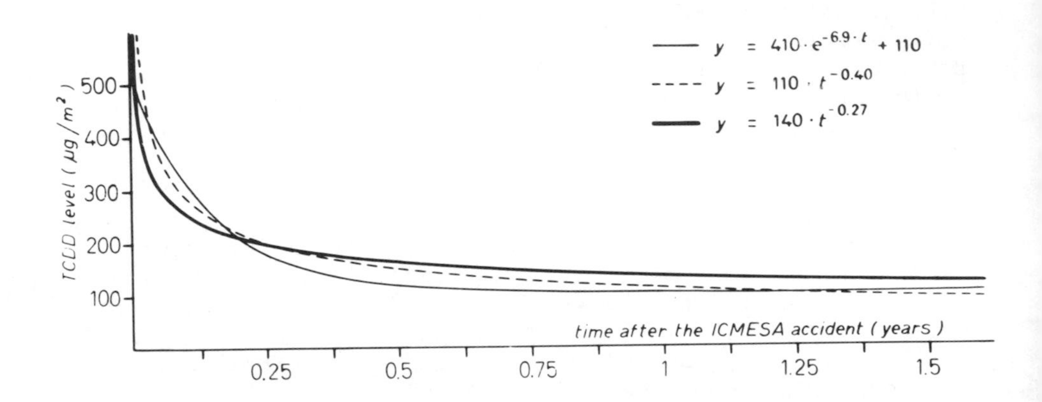

Fig. 4. TCDD persistence in the soil of Zone A may be described by the regression functions indicated. They were obtained by fitting equation $\ln \underline{y} = \underline{z} \ln \underline{t}$ *to a set of data obtained from 44 sampling sites (——) and to a 28-site data subset (---), and equation* $\ln \underline{y} = \ln(\underline{y}_0\ \underline{e}^{-kt} + \underline{y}_{00})$ *to the 28-site data subset (——) (di Domenico et al., in press). Log-transformation of TCDD values has been utilized in order to obtain Gaussian statistical distributions (di Domenico et al., 1980b). All curves appear to have a steep downslope within 0.5 years from the origin. They then flatten out at TCDD levels of approximately 100 and 150 µg/m²; TCDD levels decreased rapidly shortly after the ICMESA accident but the vanishing rate, over the following year or so, decreased so much as to produce almost negligible changes in TCDD levels in the soil.*

as high as those presently being found in the soil of Zone A.

Some highly contaminated soil samples from Zone A were also analyzed with a high resolution GS-MS system to establish whether isomers of 2,3,7,8-TCDD were present (Buser, 1977; Buser and Rappe, 1978, 1980). The data reported indicate that 2,3,7,8-TCDD accounted for over 90% of total TCDDs and that 1,3,7,8-TCDD was the major isomer present. A similar isomeric composition was also ascertained in several samples from different points inside the ICMESA reactor where the accident took place (unpublished data).

Investigations on vertical gradients were carried out in Zone A, where higher TCDD concentrations allowed detectable recoveries from greater depths (di Domenico et al., 1980c).

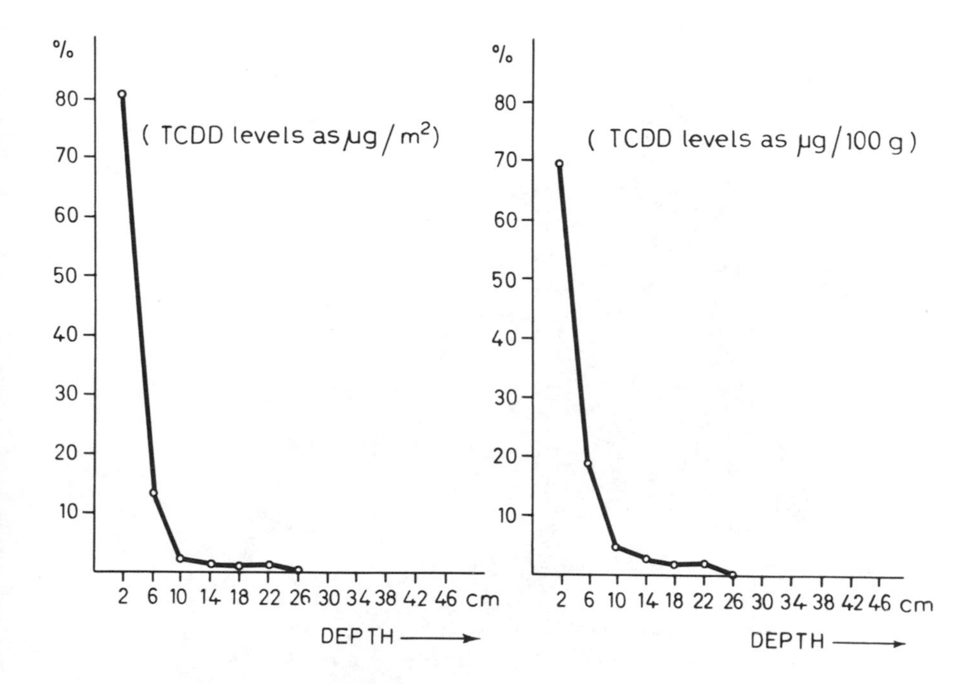

Fig. 5. Example of TCDD vertical concentration gradient in the top soil layer of a Zone A site (sampling carried out in September 1976). The results from 1976 and 1977 indicated that vertical gradients in the top soil layers appeared to vary only slightly with time. In general, over 90% of detected TCDD was found in the 15 cm-thick upper layer.

Results from 1976 and 1977 samples indicate that vertical gradients, in the deeper soil layers, varied only slightly with time. In both investigations, over 90% of the detectable TCDD was found in the upper 15 cm-thick layer (Fig. 5). As compared to findings obtained in 1976 for soil layers deeper than 8 cm slight, but significantly higher levels of TCDD were detected in 1977 (di Domenico et al., 1980c).

B. *TCDD LEVELS IN THE SOIL OF ZONE R*

Analytical results of the 1976/1977 survey for the Zones B and R are shown in Fig. 6a. TCDD levels in Zones B and R were, in general, considerably lower than those in Zone A. In fact, most TCDD levels were lower than 50 $\mu g/m^2$ in Zone B and 5 $\mu g/m^2$ in Zone R. In 1980, a large part of Zone R was re-

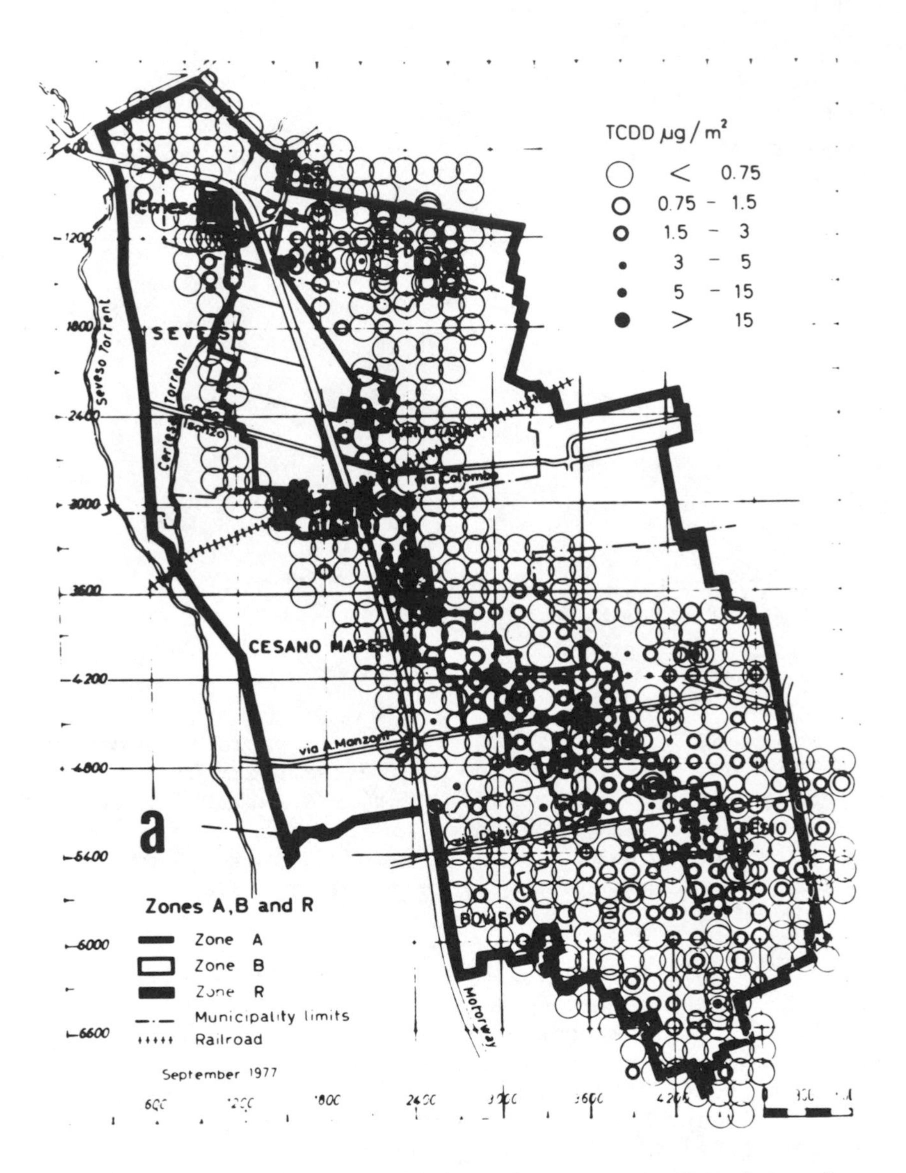

Fig. 6. September 1977 map of Zones B and R (a). Approximately 80% of data points exhibited had already been assayed by March 1977. Sampling sites were established on a 150 m square reference grid oriented north-south (as shown). Zone R soil was extensively monitored again in 1980 (b) to assess the effect of ploughing and other land treatments on TCDD concentration. Asterisks indicate sites where TCDD

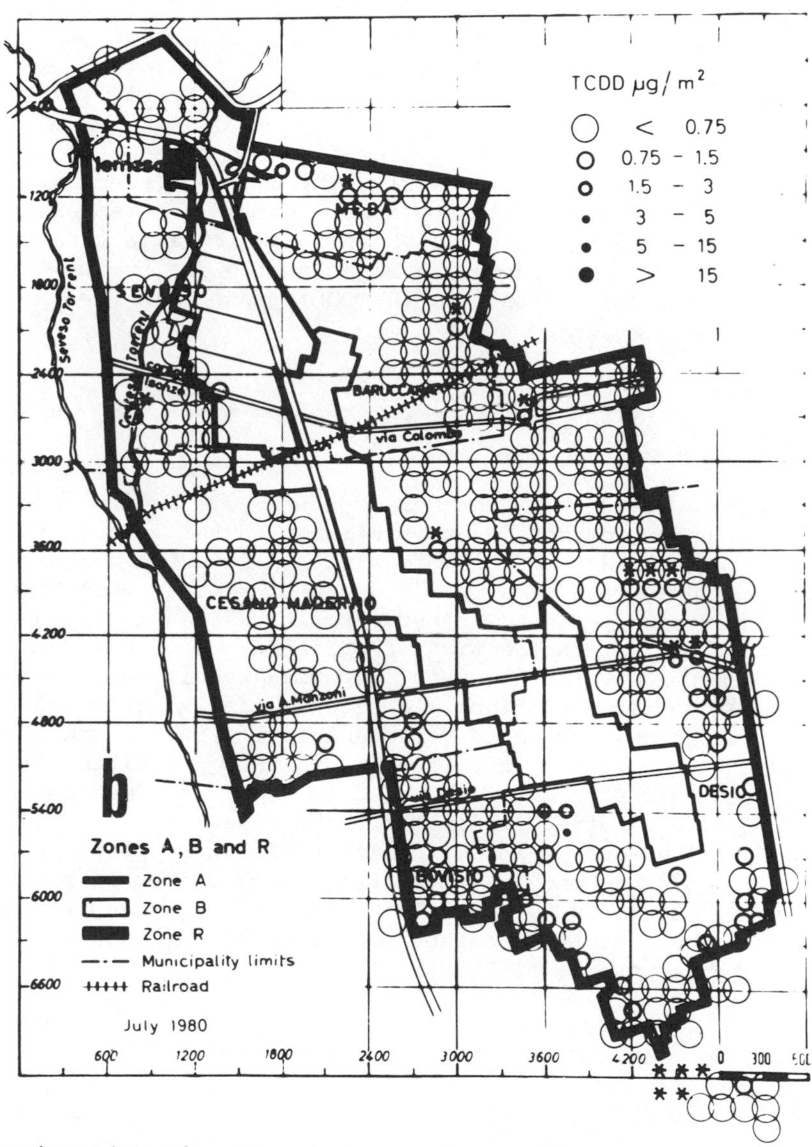

(Continuation Fig. 6)

isomers other than 2,3,7,8-TCDD were detected. In both maps sampling sites may be localized within ± 75 m of their effective locations.

monitored to evaluate the persistence of TCDD in the soil. This zone had been ploughed and worked since 1978 (Fig. 6b). A comparison of these two figures, as well as a statistical elaboration of the relevant data (Table I), indicate a significant decrease (40%) in the geometric mean level of TCDD in the soil of Zone R.

In 1980 and 1981, soil samples from ten sites of Zone R and five sites outside Zone R (Fig. 6b) were analyzed using a high resolution GS-MS system to establish whether isomers of 2,3,7,8-TCDD were present. As shown in Table II, a significant percentage of TCDD decrease can be accounted for by two TCDD isomers (1,3,6,8-TCDD and 1,3,7,9-TCDD) present in the majority of the samples tested.

C. TCDD LEVELS IN GROUND AND SURFACE WATERS AND IN SEDIMENTS

Since August, 1976 a number of tests have been periodically conducted on running water streams in the affected area as far south as the River Lambro with consistently negative results. During the same period sediment samples were taken from Torrents Certesa and Seveso. Positive results of the order of 1 ppt within the first few kilometers downstream from their confluence were obtained. Negative results were obtained farther downstream. The intensive rainfalls after the accident caused the Seveso to repeatedly overflow its embankments at the point of entry into Milan, thus depositing silt on adjacent areas. Tests conducted to determine TCDD in these silts yielded negative findings for the first four floods; the fifth flood, however, yielded positive findings (ppt). Since August, 1976, the monthly determinations conducted on pipeline and ground waters have consistently yielded negative results, even when the analytical detection threshold was as low as 1 pg/l (parts per quadrillion).

D. TCDD LEVELS IN ATMOSPHERIC PARTICLES

Airborne dust was monitored in the Seveso area to evaluate the possibility that TCDD-containing particles were airborne from the contaminated zones. High-volume samplers (particles up to 100 μm) and dust-fall jars (particles larger than 10-15 μm) located as shown in Fig. 7 were used (di Domenico et al., 1980d).

The results obtained from July, 1977 to June, 1980 from sediment specimens in the dust-fall jars are of major interest

TABLE I. TCDD Levels at 97 Soil Sites in Zone R as Reported in September, 1977 and July, 1980 Maps

Site	coordinates (x-y)	TCDD (μg/m²)	
		September 1977 map	*July 1981 map*
1	0600-0900	0.80	<0.75[a]
2	1050-1350	0.17 E+1	<0.75
3	1050-1500	0.27 E+1	<0.75
4	1650-0900	0.99	0.15 E+1
5	1800-0900	0.19 E+1	<0.75
6	1800-1050	0.13 E+1	0.26 E+1
7	1800-1500	0.12 E+1	<0.75
8	1950-1050	<0.75	0.99
9	2100-1350	0.15 E+1	<0.75
10	2250-1200	0.12 E+1	0.94
11	2250-1500	0.75	<0.75
12	2250-1650	0.76	<0.75
13	2400-1200	0.22 E+1	<0.75
14	2400-4650	0.10 E+1	<0.75
15	2550-1200	0.77	0.17 E+1
16	2550-4500	0.44 E+1	<0.75
17	2550-5100	0.17 E+1	<0.75
18	2700-1200	0.98	<0.75
19	2700-1650	0.87	<0.75
20	2700-3300	0.13 E+1	<0.75
21	2700-4950	0.37 E+1	0.84
22	2850-3450	0.11 E+1	<0.75
23	2850-3600	<0.75	0.93
24	2850-5100	0.20 E+1	<0.75
25	2850-5700	0.15 E+1	0.82
26	3000-3750	0.31 E+1	<0.75
27	3150-3600	0.11 E+1	<0.75
28	3150-3750	0.14 E+1	<0.75
29	3150-5400	0.10 E+1	<0.75
30	3150-5700	0.95	<0.75
31	3150-5850	0.23 E+1	<0.75
32	3150-6000	0.18 E+1	<0.75
33	3300-3600	0.19 E+1	<0.75
34	3300-3900	0.91	<0.75
35	3300-5850	<0.75	0.11 E+1

(Table I. continued)

Site coordinates (*x-y*)		TCDD (μg/m²)	
		September 1977 map	*July 1981 map*
36	3450-3900	0.16 E+1	<0.75
37	3450-5850	0.88	<0.75
38	3600-5400	0.16 E+1	0.25 E+1
39	3600-5550	0.14 E+1	<0.75
40	3600-5700	<0.75	0.12 E+1
41	3600-6150	<0.75	0.77
42	3750-5400	<0.75	≤0.17 E+1[b]
43	3750-5550	0.10 E+1	0.38 E+1
44	3750-6450	<0.75	0.85
45	3750-6600	<0.75	0.90
46	3900-5700	0.11 E+1	<0.75
47	3900-5850	0.11 E+1	<0.75
48	3900-6600	0.22 E+1	<0.75
49	4050-4050	0.16 E+1	<0.75
50	4050-4200	0.48 E+1	<0.75
51	4050-4500	0.17 E+1	<0.75
52	4050-5850	0.20 E+1	<0.75
53	4050-6600	0.20 E+1	0.82
54	4050-6700	0.17 E+1	<0.75
55	4200-3900	<0.75	≤0.13 E+1
56	4200-4200	0.17 E+1	<0.75
57	4200-4350	0.91	<0.75
58	4200-4500	0.14 E+1	<0.75
59	4200-4650	0.47 E+1	<0.75
60	4200-6000	0.18 E+1	<0.75
61	4200-6600	0.13 E+1	<0.75
62	4200-6750	<0.75	0.90
63	4350-3750	0.98	<0.75
64	4350-3900	<0.75	≤0.13 E+1
65	4350-4050	0.19 E+1	<0.75
66	4350-4200	0.22 E+1	<0.75
67	4350-4650	0.19 E+1	<0.75
68	4350-6000	0.24 E+1	<0.75
69	4350-6600	0.19 E+1	<0.75
70	4350-6900	0.13 E+1	<0.75

(Table I continued)

Site coordinates (x-y)		TCDD (μg/m²) September 1977 map	TCDD (μg/m²) July 1981 map
71	4500-3900	<0.75	≤0.11 E+1
72	4500-4200	0.10 E+1	<0.75
73	4500-4350	<0.75	0.16 E+1
74	4500-5850	0.66 E+1	0.84
75	4500-6000	0.11 E+1	<0.75
76	4500-6150	0.14 E+1	<0.75
77	4500-6450	0.78	<0.75
78	4500-6900	0.13 E+1	<0.75
79	4650-4350	0.13 E+1	0.24 E+1
80	4650-4500	0.24 E+1	<0.75
81	4650-4650	<0.75	0.89
82	4650-4800	0.15 E+1	<0.75
83	4650-4950	0.12 E+1	<0.75
84	4650-5100	0.18 E+1	<0.75
85	4800-4200	0.15 E+1	<0.75
86	4800-4650	<0.75	0.10 E+1
87	4800-4800	0.14 E+1	<0.75
88	4800-4950	<0.75	0.87
89	4800-6150	<0.75	0.82
90	4800-6300	0.19 E+1	<0.75
91	4950-5700	0.16 E+1	0.91
92	4950-6000	<0.75	0.97
93	4950-6150	<0.75	0.77
94	4950-7050	0.93	<0.75
95	4950-7200	0.31 E+1	0.87
96	5100-5250	<0.75	0.14 E+1
97	5100-6150	0.98	0.12 E+1

[a] *Detection threshold.*

[b] *Presence of interfering signals.*

TABLE II. TCDD Isomers Identified in Soil Specimens from 10 Sites Located in Zone R (upper group) and 5 Sites external to Zones A, B, and R (lower group)

Sampling site (x-y)	*TCDD isomer* ($\mu g/m^2$)		
	2,3,7,8-TCDD	*1,3,6,8-TCDD*	*1,3,7,9-TCDD*
0750-2700	≤0.33 E-1[a]	0.96 E-1	0.75 E-1
2250-1200	0.46	0.18	0.11
2850-3600	0.98	n.d.[b]	n.d.
3000-2100	0.97	0.11	0.75 E-1
3450-2700	≤0.15	0.21	0.13
4200-3900	0.70 E-1	n.d.	n.d.
4350-3900	0.36	0.14	0.90 E-1
4500-3900	0.66	0.14	0.90 E-1
4500-4350	0.12 E+1	0.36	0.26
4650-4350	0.60	0.15	0.90 E-1
4350-7200	0.12	0.11	0.60 E-1
4350-7350	0.70 E-1	0.80 E-1	0.50 E-1
4500-7200	0.11	0.70 E-1	0.50 E-1
4500-7350	0.21	0.13	0.80 E-1
4650-7200	0.21	0.13	0.90 E-1

[a] *Uncertain identification.*

[b] *Below detection threshold.*

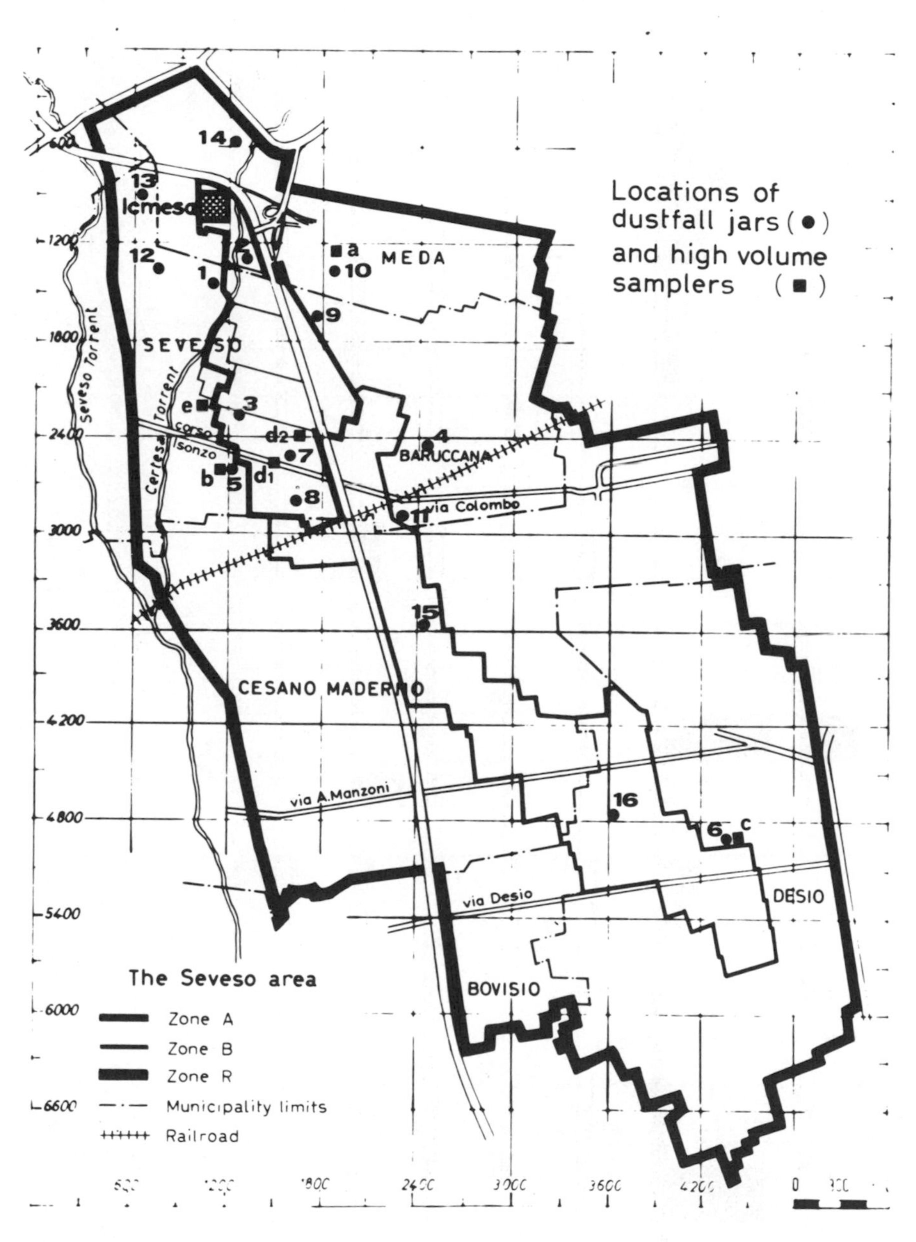

Fig. 7. Locations of dust-fall jars and high-volume samplers in Zones A, B, and R. Dust-fall jar 2 was moved 0.3 km south in June 1980.

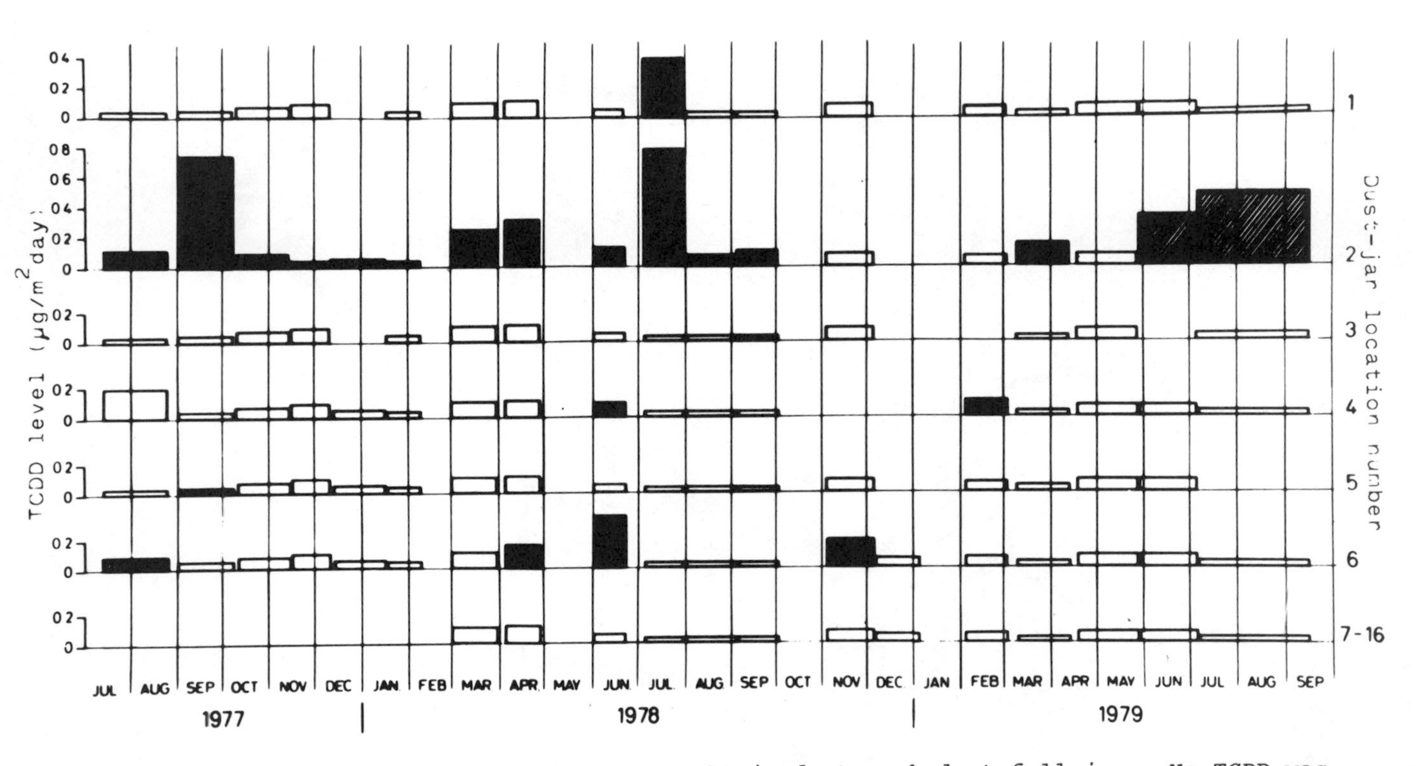

Fig. 8. Water and sediment specimens were obtained at each dust-fall jar. No TCDD was generally detected in the aqueous phase. TCDD levels in settled atmospheric particles were collected in dust-fall jars (cf. Fig. 7): hatched square, positive findings; solid square, uncertain assay; unfilled square, analytical detection threshold.

(Fig. 8). TCDD was found in fifteen out of nineteen sediment specimens sampled at Site 2 in subzone A1.

This sampling site showed seasonal variations in both dust deposits and TCDD levels. It appeared that the maximum dust and TCDD fallout deposits occurred at the beginning of the summer. TCDD levels in the settling dust also changed according to the season, although with differing patterns. TCDD levels were occasionally recorded at four other sites during other months of the year. In order to improve detection capabilities, extracts from specimens obtained from six consecutive samplings at each location were pooled after individual analysis. Results from the pooled specimens showed good agreement with those from individual specimens (di Domenico et al., 1980d). Further analyses performed from September, 1979 until June, 1980 confirmed these results. A comparison of TCDD levels in dust and soil specimens sampled at the same locations indicated that airborne TCDD did not significantly affect TCDD levels in the top soil layer. Moreover, data from dust-fall jars showed that TCDD levels in the dust decreased with increasing distance from subzone A1, moving along the main diffusion pathway of the toxic cloud. However, dust-fall jar 6 was quite anomalous in this aspect.

E. TCDD LEVELS IN PLANT TISSUES

TCDD in samples of vegetation from Zone A was assayed for the first time shortly after the accident. Samples were mainly obtained from leaves, grass, and vegetables. Samples were clearly affected by the chemical action of the cloud, indicated by withering, burns, and yellow spots. TCDD levels rapidly decreased as the distance from south-southeast from the ICMESA plant increased (Fig. 9). As shown in Fig. 9, TCDD levels in test samples ranged between 50 (approximately 175 m from the ICMESA plant) and 0.01 ppm (approximately 2000 m).

Further tests on vegetation samples from Zone A were performed by Cocucci et al. on parts that were present at the time of the accident and parts which had developed afterward (Cocucci et al. 1979). Both preexisting tissues and new growth were found to contain TCDD. Furthermore, in all tested vegetation samples the aerial parts appeared to contain higher TCDD levels than underground organs, which were, in general, somewhat lower than, or at least comparable with, that found in the soil. Parts (e.g. leaves, twigs, fruit, and bark) of several species of trees were also analyzed for TCDD. According to Coccucci et al. (1979) TCDD distribution in organs and tissues was nonrandom; TCDD levels appeared to be higher correspondingly in conductive tissues.

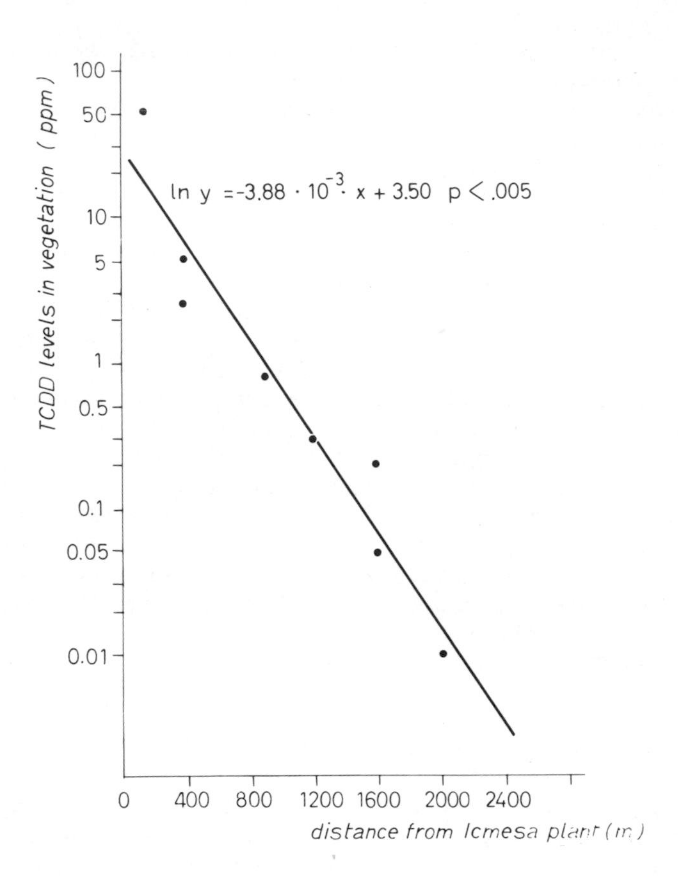

Fig. 9. Upper Zone A map from August, 1976 (Fig. 2a). Regression line of TCDD levels on vegetation versus distance from ICMESA plant obtained by fitting equation $\underline{\ln y} = \underline{b}\ \underline{x} + \underline{a}$ to TCDD analytical findings from specimens collected before July 26, 1976.

A study on the time course of TCDD levels in the epigeal and hypogeal parts of a number of cultivated species in experimental plots in Zone A confirmed the ability of plants to absorb and translocate TCDD (Cocucci, unpublished report). Epigeal and hypogeal tissues from a number of different species collected during emergence, early blooming, and at maturity generally showed a progressive increase in TCDD levels with plant maturation. An analysis of the data obtained with oats, sorghum, and darnel is reported in Fig. 10. The extent of the difference between TCDD levels in aerial and underground tissues largely depended on the stage of development of the plant. Available data suggest that TCDD levels in above-

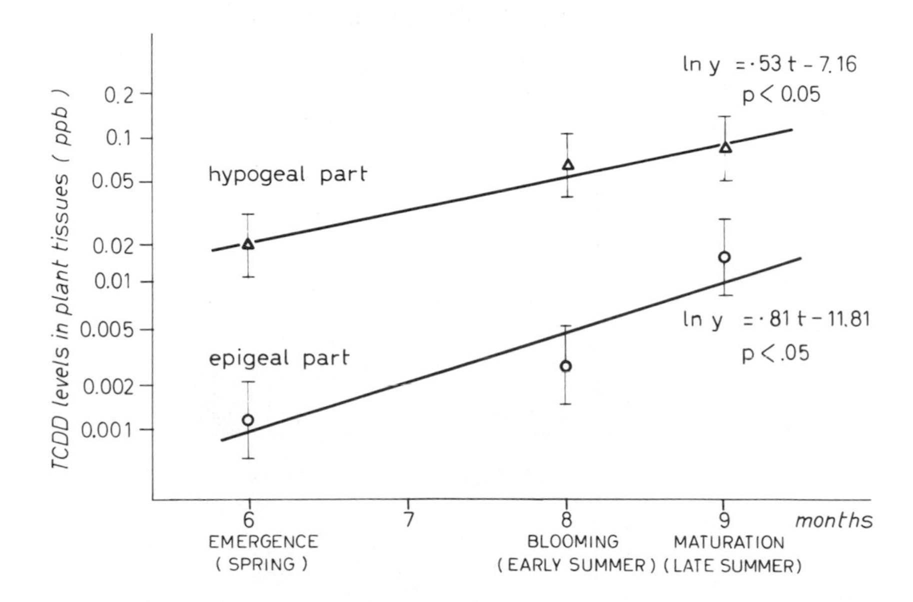

Fig. 10. Soil-to-plant TCDD translocation. The epigeal plant tissues appear characterized by TCDD concentrations generally lower than those present in the hypogeal parts. Both show increasing trends with elapsing time. Regression lines were obtained by fitting equation $\underline{\ln y} = \underline{b}\ \underline{x} + \underline{a}$ to TCDD findings in plant tissues of oats, sorghum, and darnel.

ground plant tissues were, on an average, at least 10 times lower than those found in corresponding cultivation soil.

From 1978 to 1981, epigeal and hypogeal samples of several vegetable species (onion, lettuce, carrot, potato, and small radishes) were tested for contamination by 2,3,7,8-TCDD and some isomers (Fig. 11A). Of these, 167 samples did not contain detectable 2,3,7,8-TCDD levels (detection threshold 0.2 ppt), two samples contained 2,3,7,8-TCDD levels close to 1 ppt, and thirteen samples gave uncertain results, indicating a possible, although not univocal, presence of 2,3,7,8-TCDD in the 0.4-0.8 ppt range. Last, TCDD levels in the remaining nine samples could not be determined due to interfering substances. In quite a large number of the tested samples two isomers (1,3,6,8-TCDD and 1,3,7,9-TCDD) were identified in amounts varying between 0.4 and 2.0 ppt (Fig. 11B). In particular, out of 35 epigeal vegetable samples, 8 were in the 1.2-2.4 ppt range and nine gave uncertain results indicating a possible, although not univocal, presence of these two isomers in the 0.4-1.0 ppt range.

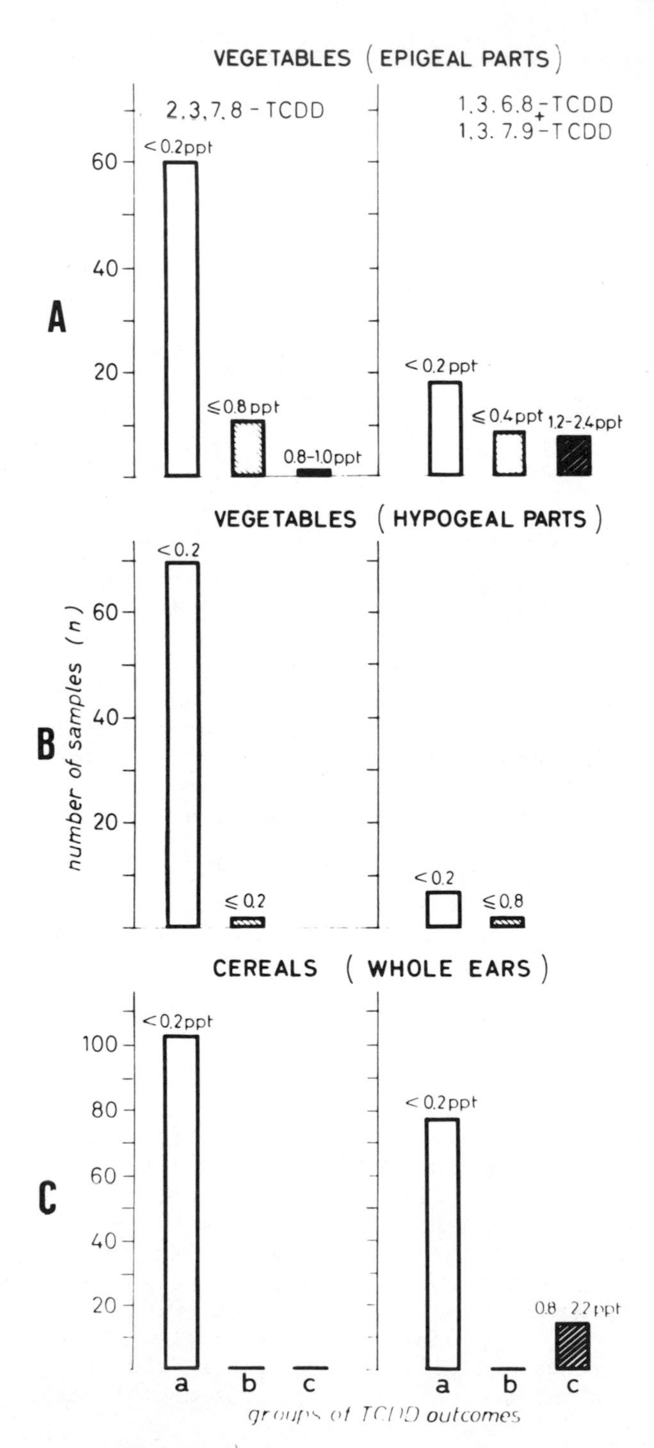
VEGETABLES (EPIGEAL PARTS)
2,3,7,8-TCDD
1,3,6,8-TCDD + 1,3,7,9-TCDD
A
< 0.2ppt
≤0.8 ppt
0.8–1.0ppt
< 0.2 ppt
≤0.4ppt
1.2-2.4ppt
VEGETABLES (HYPOGEAL PARTS)
B
< 0.2
≤ 0.2
< 0.2
≤ 0.8
number of samples (n)
CEREALS (WHOLE EARS)
C
< 0.2ppt
< 0.2ppt
a
b
c
a
b
c

An extensive monitoring of TCDD contamination of wheat and oats was performed in 1980 by sampling, throughout the maturation period, ears and 10-cm stems from twelve experimental plots randomly distributed in Zone R. Of 103 tested samples (Fig. 11C) not one contained detectable levels of TCDD (detection threshold 0.5 ppt); however, seventeen samples obtained from three of these plots contained detectable levels of the two isomers, namely, 1,3,6,8-TCDD (0.76-2.16 ppt) and 1,3,7,9-TCDD (0.32-1.68 ppt). After harvesting, 97 samples of wheat and rye kernels were tested for 2,3,7,8-TCDD. Of these, fifteen were also tested for the two aforementioned isomers. No TCDD was detected in any of the samples.

F. TCDD LEVELS IN FARM ANIMALS

At the time of the accident, 81,131 animals (24,885 rabbits, 55,545 poultry and other small animals, 349 cattle, 233 pigs, 49 horses, 21 sheep, and 49 goats were inhabiting Zones A, B, and R. Death of rabbits and poultry started some days after the accident and markedly increased within the first 2 weeks. By the end of August, 1976, 2062 rabbits and 1219 small farmyard animals had succumbed; no deaths were reported, at that time, among cattle, horses, pigs, sheep, and goats. Figure 12a shows the locations of farms where death occurred. Figure 12b shows the locations, in Zones B and R, of rabbits that were found to exhibit positive TCDD levels in hepatic tissue. Seventy-five percent of the farms shown in Fig. 12a were in Zone A, 22% in Zone B, and 14% in Zone R. Of the animals that died, 31.9% were in Zone A, 8.8% were in Zone B, and 6.8% were in Zone R (Veterinary Service of Lombardy Region and "Mario Negri" Institute Report, 1980). About 45% of dead rabbits were necropsied. A significant percentage was found to exhibit a pathological syndrome characterized by substernal and retrosternal edema, hemorrhagic tracheitis, pleural serous hemorrhage, and distrophic lesions of hepatic tissues. An overall view of the TCDD analyses performed on animal tissues from 1976 to 1979 is given in Table III (TCDD detection threshold 250 ppt). TCDD was de-

---

Fig. 11. Isomer presence frequency at different contamination levels of 2,3,7,8-, 1,3,6,8-, and 1,3,7,9-TCDD in epigeal (A) and hypogeal (B) parts of vegetables and in cereal tissues (C) from Zone R specimens: (a) absence (detection threshold 0.2 ppt); (b) identification uncertain; (c) identification certain.

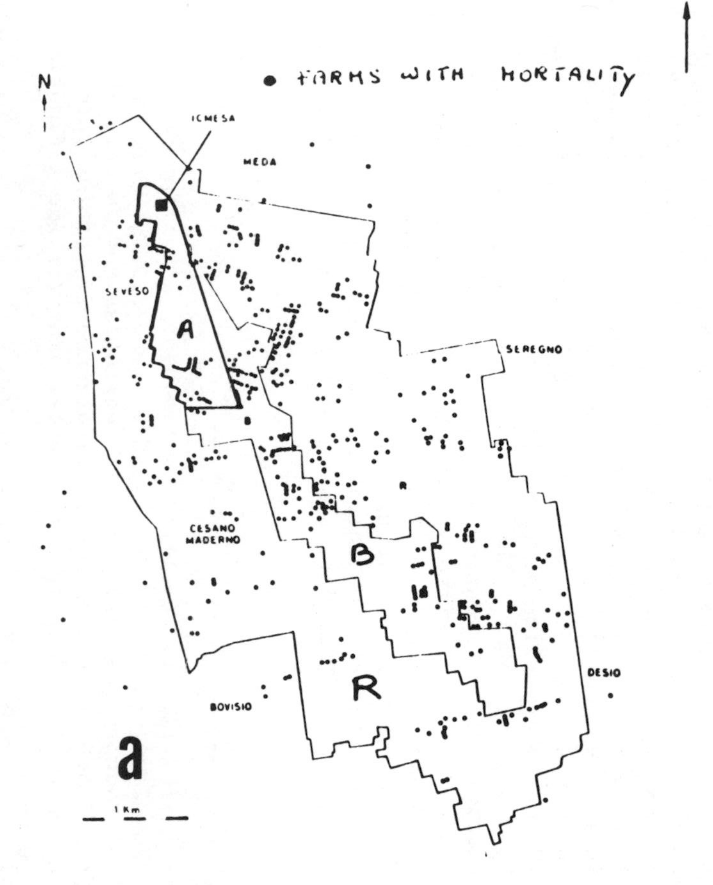
N
FARMS WITH MORTALITY
ICMESA
MEDA
SEVESO
A
SEREGNO
CESANO
MADERNO
B
R
DESIO
BOVISIO
a
1 Km

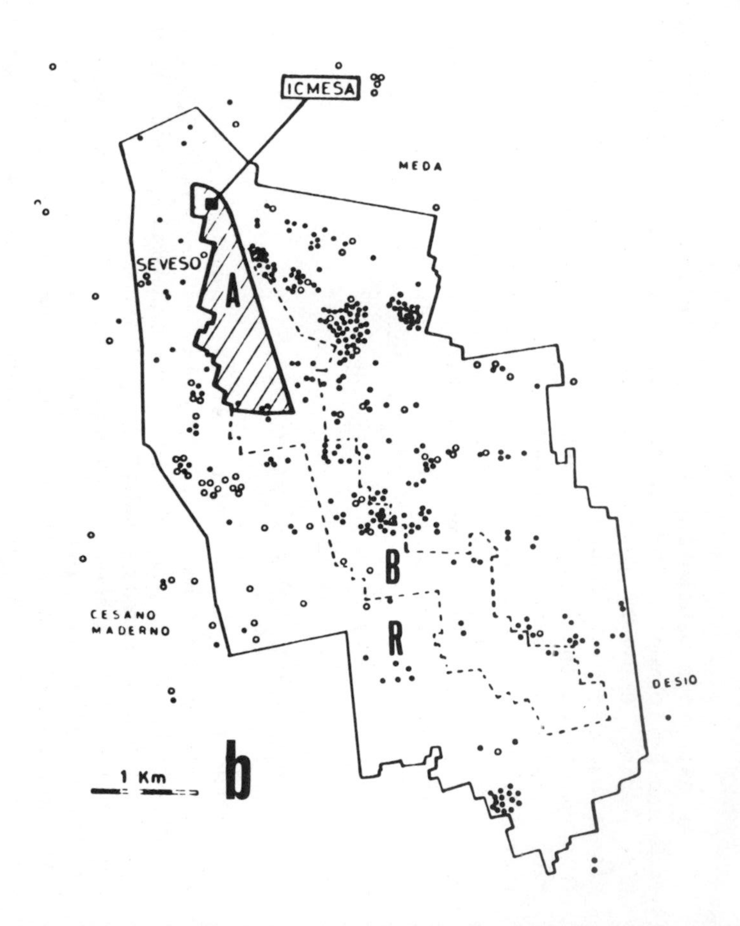
N
ICMESA
MEDA
SEVESO
A
B
R
CESANO
MADERNO
DESIO
1 Km
b

TABLE III. TCDD Analyses on Liver of Farm Animals from Contaminated Zones and Surrounding Areas (1976-1979)[a]

Animal	*Number of samples*	*TCDD-containing samples*	*TCDD maximum level (ppb)*
Rabbits[b]	698	433	633
Poultry	83	35	24
Cattle	43	21	94
Horses	12	2	88
Pigs	13	0	-
Goats	25	17	1
Cats	1	0	-

[a] *Data by Veterinary Service of Lombardy Region et al. (1980).*

[b] *Figures include rabbits kept in the special test plots on contaminated ground for experimental purposes.*

tected in 97, 92, and 75% of Zone A, B, and R specimens, respectively. The pathological syndrome was not always found in animals that had positive levels of TCDD in the liver.

G. TCDD LEVELS IN COWS' MILK

About 2 weeks after the accident, cows from Zone A were transferred to a special cowshed under sanitary surveillance. Cattle breeding in Zones B and R were fed fodder from distant areas. Milk samples were collected shortly after the accident (July 27-August 28, 1976) from cattle in the contaminated zones and in surrounding areas. These samples usually consisted of pooled milk from the individual cows from each farm. TCDD levels in the milk ranged from less than 40 (sensitivity threshold) to about 7900 ng/liter (Fanelli et al., 1980). The re-

Fig. 12. The map on the left (a) shows the farm locations in Zones A, B, and R (●) where animal deaths were reported from July 10 to August 31, 1976. The map on the right (b) shows the locations where tested rabbit livers yielded positive (●) or negative (○) TCDD findings in 1976.

sults obtained showed a wide variation in TCDD levels according to farm location, higher levels being found in milk samples from the farms closer to the ICMESA plant. At that time, a few contaminated milk samples were also obtained from outside Zone R. An intensive monitoring of cows' milk from the whole area surrounding Zone R was begun in March, 1978. Twenty-eight farms were checked repeatedly. Cows from only one farm from Varedo (south of Zone R) exhibited detectable TCDD levels (range 20-32 ng/liter of milk). It was later found that these cows had been fed in part from fodder harvested in Zone R near Desio. Thereafter, these cows were fed fodder purchased only in uncontaminated areas. After a few months, the TCDD levels in milk became undetectable.

H. TCDD LEVELS IN WILDLIFE ANIMALS

An overall view of the TCDD analyses performed from 1976 to 1979 on wildlife animals from the contaminated zones and surrounding areas is shown in Table IV.

TABLE IV. TCDD Analyses on Wildlife Animals from Contaminated Zones and Surrounding Areas (1976-1979)[a]

Animal	*Tested organs and number of samples*	*TCDD-containing samples*	*Maximum level of TCDD*
Hares	6 (Liver)	4	13
Field mice	14 (Whole body)	14	49
Rats	1 (Pool 4 livers)		28
Earthworms	2 (Pool)		12
Frogs	1 (Liver)		0.2
Snakes	1 (Liver)		3

[a]*Data by Veterinary Service of Lombardy Region et al. (1980).*

Further research on TCDD levels in small young wild mammals from Zone A and on earthworms from Zones A, B, R, and surrounding areas was carried out by Omodeo and co-workers from November, 1979 to March, 1981. They were able to detect TCDD in both skin and liver of a few specimens of *Apodemus sylvaticus* and in the liver of one specimen of *Rattus norvegicus*. TCDD levels in the skin ranged from 0.4 to 0.7 ppb and in the liver from 0.4 to 31.4 ppb. The extensive work performed by Omodeo and co-workers has shown that earthworms ingesting TCDD-containing soil tended to concentrate TCDD in their tissues approximately ten times more than levels found in the soil and to excrete the unabsorbed TCDD at the soil surface. Moreover, TCDD levels did not affect the survival of earthworms present in Zones A, B, and R. However, moles and other insectivorous animals have apparently disappeared from Zone A.

IV. DISCUSSION

The synthesis of trichlorophenol (TCP) at the Givaudan-La Roche ICMESA plant, involved the conversion of 1,2,4,5-tetrachlorobenzene to TCP by alkaline hydrolysis at approximately 160°C. Trace amounts of TCDD were also produced under standard operating conditions. On July 10th, 1976, however, the exothermic process begun in the reaction bulk resulted in the overproduction of TCDD. The safety valve pressure disk connected to the reactor ruptured and the fluid mixture burst through the valve into the air. The visible portion rose approximately 50 m and subsequently fell back to earth. Moving away from the main diffusion pathway, and/or from the ICMESA plant, a gradual dilution process occurred within the chemical cloud because of its three-dimensional airborne motion and simultaneous loss of material. The settling of the cloud resulted in the contamination of all the exposed surfaces. A slight (few centimeters) penetration of TCDD into the soil then occurred. Because TCDD is only slightly soluble in water, its limited vertical migration in the soil may have occurred by binding to soil colloids and particles. The total amount of TCDD recovered from the soil in the affected areas was estimated on the order of several hundreds of grams. Clearly, TCDD recovered in soil represented only a small fraction of the total amount of TCDD released into the environment as a consequence of the ICMESA accident.

Extensive monitoring throughout the affected area has shown several interesting features of the environmental fate of TCDD. The available data provide statistically significant evidence that TCDD levels dropped in the unworked soil of Zone A within 5 months of the accident. Following this period, no further decrease in TCDD levels was detected. The pattern of TCDD abatement in the soil of Zone A is consistent with mathematical functions suggesting that TCDD disappearance rate decreased with time. One month after the accident, the apparent mean TCDD half-life was approximately 10-14 months; 17 months after the accident, the apparent half-life was estimated to be more than 10 years. In the soil of Zone R, ploughed and worked since 1978, TCDD levels in surface soil decreased by about 40% in 3 years (1976/1977-1979/1980). Findings concerning the vertical and surface distribution of TCDD in soil at different times after the ICMESA accident are consistent with the hypothesis that the initial abatement of TCDD observed in Zone A was at least partially the result of photodegradation which may have occurred in the topmost soil layer when most TCDD was exposed to direct sunlight and partially due to TCDD volatilization through different channels, the extent of which may have decreased with time. It is relevant to note that TCDD was found to be more stably bound to the soil of Zone A than TCDD which was added as an iso-octane solution to samples of the same type of soil (Viviano et al., 1979). This may explain the TCDD abatement observed in Zone R, where ploughing has effectively mixed the soil. Another factor which might have contributed to TCDD abatement is the capability of cultivated plant species to absorb TCDD from soil. Levels of TCDD in vegetation were generally somewhat lower than, or at most, comparable to those levels found in the soil surface. Earthworms have been found capable of concentrating TCDD in their bodies to levels approximately 10 times higher than soil levels. Similarly, farm and small wildlife mammals have been observed to accumulate TCDD in their livers (however, a definite relationship between TCDD levels in soil and mammal liver has not yet been established). Of all tested environmental systems, the only one that has been consistently found to be TCDD-free is ground and surface waters, which may be due to very low solubility of TCDD in water and its corresponding high affinity for soil and sediments. The comparison of TCDD levels in dust and soil samples taken at the same locations indicates that airborne TCDD did not significantly affect TCDD levels in the soil top layer.

Today, some 5 years after the accident, the extreme seriousness of the environmental impact of the ICMESA accident contamination can be appreciated. Zone A, evacuated in 1976, is now a site of deep environmental degradation. Utilization of locally grown vegetables and animal breeding has been prohibited and construction has been restricted in Zones B and R. By the end of August, 1976, 14,300 small farmyar animals had been slaughtered in Zones A and B. By the end of January, 1977, 51,000 small animals had been slaughtered in Zone R.

The slaughter of 277 cattle, 30 horses, 50 sheep and goats, and 44 pigs occurred by the end of December, 1978. Not only was the whole farm animal population of Zones A, B, and R destroyed as a consequence of the ICMESA accident, but wildlife animals were also severely contaminated.

One factor that has contributed to the impact of the ICMESA accident on the environment is the very high persistence of TCDD in the soil and the apparent lack of any suitable method to inactivate it. Because the vertical distribution of TCDD in soil is unusual, stripping the top 25 cm-thick layer of soil would provide an average removal of more than 90% of the total TCDD. Thoroughly mixing the top 30 cm-thick layer produces a drop in TCDD levels in the upper 7 cm-thick soil layer ranging from one-half to one-third, depending on the effectiveness of the treatment. Because of these factors, the reclamation process of Zone A has largely been based on removing top soil layers and transferring the soil to underground containers located in the same zone. It is evident that, while this reclamation process reduces the risk for the population and prevents any further environmental degradation of Zone A, it does not allow reestablishment of the conditions which existed prior to the accident.

The extensive environmental monitoring performed as a consequence of the accident has shown the presence of traces of two isomers of 2,3,7,8-TCDD in several soil and vegetation samples of Zone R. It has also suggested the possible existence of additional contamination sources in the Seveso area. The significance of these findings is now under investigation and the local municipal incinerators as well as other combustion systems operation in the zone are being evaluated as possible contributing contamination sources.

Last, it should be noted that the ICMESA accident caused one of the most serious accidental emergencies ever produced, by releasing into the environment toxic chemicals. One feature of this emergency was the apparent lack of immediate understanding of the nature and seriousness of the emergency, as well as of the extent of the area which was affected. This became clear only after the progressive appearance of adverse effects on both the inhabitants and the environment. It also became clear that six municipalities of the Lombardy Region in the province of Milan had been affected and, for safety purposes, the Authorities were forced to place the inhabitants of eleven municipalities (about 220,000 persons) under medical and epidemiological surveillance. Therefore, the Seveso emergency had an interjurisdictional character--the focal point of the response system had to be at a regional level. About 16 days after the accident, when the Regional Minister for Health was invested with full powers by the Lombardy Regional

Council, there was an official response. The Lombardy Region has since been responsible for all relevant measures, utilizing the resources made available at the regional level by a Government allocation of about 100 million U.S. dollars and employing the resources of the municipalities, provinces, and industries involved.

ACKNOWLEDGMENTS

We wish to express our great appreciation to the Lombardy Region Authorities for their collaboration. We are also grateful to G. Piva Micozzi for her technical help in the preparation of the original drawings, A. Lezza for his photographic assistance, and S. L. Galantini, M. Dell'Aquila, and L. Sansone for their technical assistance in the typing of the text.

REFERENÇES

Adamoli, P., et al. (1978). Analysis of 2,3,7,8-tetrachlorodibenzo-*p*-dioxin in the Seveso area. *Ecol. Bull. 27,* 31-38.

Buser, H. R. (1977). Determination of 2,3,7,8-tetrachlorodibenzo-*p*-dioxin in environmental samples by high resolution gas chromatography and low resolution mass spectrometry, *Anal. Chem. 49,* 918-922.

Buser, H. R., and Rappe, C. (1978). Identification of substitution patterns in polychlorinated dibenzo-*p*-dioxins (PCDDS) by mass spectrometry. *Chemosphere 7,* 199-211.

Buser, H. R., and Rappe, C. (1980). High resolution gas chromatography of the 22 tetrachlorodibenzo-*p*-dioxin isomers, *Anal. Chem. 52,* 2257-2262.

Cavallaro, A., Bartolozzi, G., Carreri, D., Bandi, G., Luciani, L., Villa, G., Gorni, A., and Invernizzi, G. (1980). A method for the determination of 2,3,7,8-tetrachlorodibenzo-*p*-dioxin at ppt levels in vegetables by high-resolution gas chromatography and low-resolution mass spectrometry, *Chemosphere 9,* 623-628.

Cavallaro, A., Bandi, G., Mangini, E., Invernizzi, G., Luciani, L., and Gorni, A. (1983). Negative ion chemical ionization mass spectrometry as a structural tool in the determination of small amounts of PCDDs and PDDFs in environmental samples. *Proc. Workshop Impact Chlorinated Dioxins and Related Compounds on the Environment, Rome, October 22-24, 1980,* in press.

di Domenico, A., Silano, V., Viviano, G., and Zapponi, G. (1980a). Accidental release of 2,3,7,8-tetrachlorodibenzo-

p-dioxin (TCDD) at Seveso, Italy: I. Sensitivity and specificity of analytical procedures adopted for TCDD assay. *Ecotoxicol. Environ. Safety 4,* 283-297.

di Domenico, A., Silano, V., Viviano, G., and Zapponi, G. (1980b). Accidental release of 2,3,7,8-tetrachlorodibenzo-*p*-dioxin (TCDD) at Seveso, Italy: II. TCDD distribution in the soil surface layer. *Ecotoxicol. Environ. Safety 4,* 298-320.

di Domenico, A., Silano, V., Viviano, G., and Zapponi, G. (1980c). Accidental release of 2,3,7,8-tetrachlorodibenzo-*p*-dioxin (TCDD) at Seveso, Italy: IV. Vertical distribution of TCDD in soil. *Ecotoxicol. Environ. Safety 4,* 327-338.

di Domenico, A., Silano, V., Viviano, G., and Zapponi, G. (1980d). Accidental release of 2,3,7,8-tetrachlorodibenzo-*p*-dioxin (TCDD) at Seveso, Italy: VI. TCDD levels in atmospheric particles. *Ecotoxicol. Environ. Safety, 4,* 346-356.

di Domenico, A., Viviano, G., and Zapponi, G. (1983). Environmental persistence of 2,3,7,8-TCDD at Seveso. *Proc. Workshop on Impact of Chlorinated Dioxins and Related Compounds on the Environment, Rome,* October 22-24, 1980, in press.

Fanelli, R., Bertoni, M. P., Bonfanti, M., Castelli, M. G., Chiabrando, C., Martelli, G. P., Noè, M. A., Noseda, A., and Sbarra, C. (1980a). Routine analysis of 2,3,7,8-tetrachlorodibenzo-*p*-dioxin in biological samples from the contaminated area of Seveso, Italy. *Bull. Environ. Contam. 24,* 818-823.

Fanelli, R., Bertoni, M. P., Bonfanti, M., Castelli, M. G., Chiabrando, C., Martelli, G. P., Noè, M. A., Noseda, A., Garattini, S., Binaghi, C., Marazza, V., Pezza, F., Pozzoli, D., and Cicognetti, G. (1980b). 2,3,7,8-Tetrachlorodibenzo-*p*-dioxin levels in cow's milk from the contaminated area of Seveso, Italy. *Bull. Environ. Contam. Toxicol. 24,* 634-639.

Viviano, G., Zapponi, G., di Domenico, A., Marinelli, A., and Bianchi, E. (1979). Metodi di rilevamento della TCDD veicolata dal particolato atmosferico nell'area di Seveso. *Ann. Ist. Super. Sanità 15,* 719-736.

DISCUSSION

DR. TUCHMANN-DUPLESSIS: You mentioned that the soil contained TCDD and I wonder about the value of the earthworms you discussed in the soil as an index of the degree of TCDD contamination.

DR. KEARNY: This organism is resistant to TCDD and will withstand approximately 10 times the amount of exposure that is lethal to other species. This unique behavior has been studied in agriculture departments and may be useful for the study of various kinds of soil contamination. The progeny are also resistant.

DR. AYRES: Plant tissue fixation is also suggested as a possible source of data on the level of exposure to TCDD, especially differences in amounts on leaf surfaces and inner tissue cells.

DR. POCCHIARI: The amounts taken up by plants is extremely small, in general, but measurable amounts of TCDD can be found.

DR. REGGIANI: A year ago there was a workshop on the toxicity of halogenated compounds and TCDD. The general review of the many variables and unknowns and the toxic symptoms, especially the chronic effects of low levels of all chemicals, have now become a public concern in general. I want to give you now a report on what we know about these things at this time.

DR. COULSTON: Dr. Reggiani succeeded in giving many novel and useful health protective measures during the period of great concern following the Seveso incident.

ISBN 0-12-193160-9

CHAPTER 2

AN OVERVIEW ON THE HEALTH EFFECTS OF HALOGENATED DIOXINS AND RELATED COMPOUNDS--THE YUSHO AND TAIWAN EPISODES

G. Reggiani

F. Hoffmann-La Roche & Co. Ltd.
Basel, Switzerland

I. INTRODUCTION

There are numerous health effects and symptoms and signs reported in humans following the exposure to halogenated dioxins and related compounds. They are given in the tabulation below and have been adapted after Kimbrough (1).

System or finding	*Signs of lesion*	*Chemical product*
General	Weakness, headache, weight loss, impotence, insomnia, loss of appetite, nausea, abdominal pain	Technical PCP, polychlorinated naphthalenes, -dibenzodioxins, -dibenzofurans, -biphenyls, polybrominated biphenyls, dibenzofurans

ISBN 0-12-193160-9

System or finding	*Signs of lesion*	*Chemical product*
Skin	Chlor- or bromacne, hyperpigmentation, hair loss, nail growth, porphyria	Penta- and hexachloronaphthalenes, tetrachlorodibenzodioxins, tri- and tetrachlorodibenzofuran, hexa- and heptachlorobiphenyls, tetrabromodibenzofuran, polybrominated naphthalene
Hepatic	Liver enlargement, impaired liver function, fatty degeneration, fibrosis, hepatocellular necrosis	Chlorinated naphthalenes, tetrachlorodibenzodioxin, technical pentachlorophenol
Respiratory system	Reduced vital capacity, chronic bronchitis	Technical pentachlorophenol, polychlorinated and polybrominated biphenyls
Central and peripheral nervous system	Muscle and joint pain, numbness and spasm in the limbs, reduced sensory and motor conduction velocity, sensory disturbances (taste, hearing, smell, sight impairment)	Polychlorinated biphenyls, -dibenzofurans, -dibenzodioxin
Endocrine system	Irregular menstrual cycle, retarded growth	Polychlorinated biphenyls and -dibenzofurans
Ocular manifestations	Several degrees of inflammation with burning, discharge, edema of eyelid, cyst formation of tarsal glands and pigmentation of eyelid and conjunctiva	Technical pentachlorophenol, polychlorinated biphenyls, -dibenzofurans and -dibenzodioxin

System of finding	Signs of lesion	Chemical product
Stomatological alterations	Pigmentation of oral mucosa, anomalies in number and growth of teeth	Polychlorinated biphenyls and -dibenzofurans
Clinical chemistry laboratory findings	Increased serum triglycerides, decreased serum bilirubin	Polychlorinated biphenyls, -dibenzofurans and -dibenzodioxins

The clinical features have been observed mainly in cases of occupational exposure where dosage is unknown, route of administration is assumed, latency period to onset of toxic effects is poorly defined, length and intensity of exposure is seldom recorded or available, number of cases entering the study is imposed by circumstances beyond the investigators, baseline data for control are often not adequate, and a comparison with a matching sample is lacking. In addition, the sample consisted mainly of adult males.

The majority of the cases of exposure reported have been cases of serious toxicity, i.e., of unquestionable toxic manifestations which assume exposure and, consequently, absorption of many times the threshold dose level to produce toxic effects. Furthermore, these have been cases of mainly acute exposure following a change in a manufacturing procedure or a lack of safety measures or accidental exposure.

These health effects might somewhat provide indications for long-term exposure to low doses of the same chemicals or for their effects when stored in the body in trace quantities.

Minute amounts of many chemicals can be detected in the human body using very sensitive analytical chemical methods as these methods improve (2) the detection of very low quantities of a chemical in body fluid should provide additional data on adsorption (Table I) and consequently exposure.

II. DETERMINATION OF EXPOSURE

Contact with one or several chemicals assumes an exposure to and consequent absorption of these chemicals. However, the assessment of the level of exposure and of the quantity of the actual absorption are not as easy to ascertain. Terms such as high, medium, and low exposure (dosage exceeding background

TABLE I. Trace Analysis and Its Implications (Tetrachlorodibenzodioxins)[a]

Amount	*Weight (gm)*	*Trace contamination level per gm*	*Number of molecules*	*Level of analytical expertise required to detect residues*
Milligram (mg)	10^{-3}		10^{18}	Titration
Microgram (μg)	10^{-6}	1 ppm	10^{15}	Spectrophotometry
Nanogram (NG)	10^{-9}	1 ppb	10^{12}	Gas chromatography
Picogram (pg)	10^{-12}	1 ppt	10^{9}	Mass spectrometry
Femtogram (fg)	10^{-15}	1 ppq	10^{6}	Mass spectrometry
Attogram (ag)	10^{-18}		10^{3}	Mass spectrometry
Mologram ?	10^{-21}		(1)	Mass spectrometry

[a] *After Cairns et al., 1980 (2).*

exposure) are all relative and vary quite widely as noted in the literature.

Of the clinical signs and symptoms of toxicity, only chloracne is sufficiently specific for and, therefore, indicative of exposure (3). However, not all chemicals are chloracnegenic and those which are have different levels of acnegenicity. Furthermore, the skin lesions tend to be spontaneous and completely heal. Thus, they may have occurred and disappeared and are not observable at the time of clinical examination. In addition, in its mildest form (a few blackheads on the face around the malar region), it can escape detection by the examining clinician. Finally, the organ susceptibility in man may differ from individual to individual, thus determining whether one develops other toxic effects, irrespective of chloracne.

The detection of the chemical(s) in body fluids and tissues will provide a new dimension in the assessment of exposure i.e., the process of uptake of the chemical from the environment into the systemic circulation. Given the parameters exposure, the detection of the chemical in the blood and urine will provide the evidence that the chemical was not only available but also absorbed. The improved sensitivity of technical analysis will also open the way to the study of the kinetics of the chemical in the human body. Thus, time and level of exposure can be calculated and correlated with the clinical signs and symptoms of toxicity.

III. RISK ASSESSMENT OF LOW DOSES

Experimental data (4-9) have shown that for some chemicals effects are linked with the structure of each isomer. One isomer can induce enzyme production and acnegenic properties at very low doses, while a change in the position of one chlorine atom in the molecule in another may not produce these effects at all. With the exception of 2,3,7,8-TCDD very little is known about the toxicity of these other chemicals. In fact, for the majority of these chemicals, long-term toxicity studies and studies on their effects on reproduction are still lacking.

There is considerable concern in the medical community about the proper methodology to be used for the study of the long-term effect of exposure to low doses of these chemicals. Although the present diagnostic means are adequate for identifying subtle clinical, histologic, and biochemical changes, doubts are reaised about the ability to understand the meaning of these findings and their role in predicting the long-term effects on human health. There is uncertainty concerning whether borderline changes observed in morphologic (liver histology) and blood chemistry tests, immunological tests, cytogenetic tests, nerve conduction velocity tests, and respiratory function tests are the expression of simple physiological adjustment or an unfavorable reaction of an organ which will lead eventually to irreversible damage. In the great majority of cases the chemically induced changes in some functions or structures are probably only an indicator of an exposure to quantities of chemicals which are below the toxic level. The correlation of the level of the chemical(s) in body fluids, may provide a sensitive system that can be used to monitor exposure of humans to the chemicals and the importance of that exposure in the long-term susceptibility to disease. They are not, however, a symptom of disease. Failure to agree on the meaning of these findings masks the definition of "observable adverse health effects."

IV. HUMAN RESPONSE TO EXPOSURE

In the final analysis it is heavy exposure to chemicals that provides an opportunity of measuring human response. Extrapolation from one dose level to another might provide data for estimating the human risk associated with low exposure levels.

There have been a number of episodes of heavy exposure to chemicals. They have involved both sexes of all ages, and a sizable portion of the population. In addition, the dose, route of absorption, and length of exposure could be somewhat defined. A number of these cases are presented below.

In 1973, in Michigan, over 10,000 residents were exposed to polybrominated biphenyls (PBBs). The chemical was absorbed through the consumption of meat, milk, and other dairy products. Of this population, the level of exposure for 4000 residents was determined, although the accuracy of these results were not definitive. The PBB serum levels ranged from 0 to 1900 μg/liter but the daily and total consumption was not established (10,11). Thus far, it has not been shown conclusively that PBBs caused illness in most of the exposed population.

In 1976, in Seveso, Italy, over 5000 residents were exposed to 2,3,7,8-TCDD. For several years, these residents have lived and are still living in a territory contaminated with a known amount of the chemical. The daily and total amount of exposure and the route of absorption of this chemical are not known. However, chloracne was observed in a small subgroup of the population, although this health effect does not provide conclusive information about the route of absorption, the daily intake, the length of exposure, and the total dosage (12).

At the Love Canal in Niagara Falls, New York, 239 families were exposed to more than 200 organic chemical compounds in a landfill. These 239 families had to be relocated. Of the 22,000 tons of chemical wastes buried in the landfill, there are 200 tons of TCP with approximately 300 ppb of 2,3,7,8-TCDD (about 60 g). The conditions of exposure of this episode are exceedingly complex. The population at risk was exposed to many chemicals and one cannot say if one, several, or an interaction between them, are likely to be responsible for effects observed (13).

Approximately 1600 residents were exposed to a mixture of polychlorinated biphenyls (PCB's), polychlorinated dibenzofurans (PCDF's) and polychlorinated quaterphenyls (PCQ's) in 1968 in South Japan. These residents consumed Yusho contaminated with an edible oil (Yusho poisoning) (14-17).

A similar event occurred in 1978 in central Taiwan, where 2000 residents were exposed to the same mixture of PCB's, PCDF's and PCQ's (18,19).

In the data available, only the last two episodes provide precise information about the daily intake, length of exposure and total dosage, and latency to onset of effects and time of the appearance of the single symptoms and signs of the clinical picture of the poisoning. These episodes are discussed in greater detail in the next section.

V. THE 1968 YUSHO AND THE 1978 TAIWAN EPISODES

A sporadic outbreak of a peculiar skin disease in South Japan between March and October 1968 was the beginning of the so-called "Yusho disease." It was caused by contamination of rice bran oil by Kanechlor, a commercial brand of a polychlorinated biphenyl which leaked from a heating pipe. About 1600 people were identified as having ingested the contaminated rice oil.

On June 7, 1968 a 3-year-old girl was admitted to the Department of Dermatology of the Kyushu University with symptoms of chloracne. Her parents and her elder sister showed the same symptom, leading health authorities to become suspect of the cause of the disease. Similar cases were uncovered in the weeks and months that followed (14-17).

In Taiwan, during the last months of 1978, for reasons which still require explanation, rice cooking oil produced by a local company was contaminated with material containing PCB. In March, 1979, the first cases of unusual ocular and dermatologic manifestations were observed. They occurred mainly in the central region of Taiwan and their number increased rapidly to about 2000. In October, 1979, the Central Health

Authorities in Taiwan issued a statement announcing that the cause of the disease had been the ingestion of edible oil contaminated with PCB (18,19).

VI. COMPOSITION OF THE TOXIC MATERIAL--QUANTITY AND DURATION OF INGESTION

The toxic chemicals involved in the two episodes discussed in the previous sections (20-25) were polychlorinated biphenyls (PCB's), polychlorinated dibenzofurans (PCDF's) and polychlorinated quaterphenyls (PCQ's). Their chemical structures are presented below.

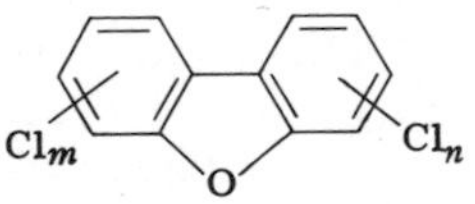

Cl_m O Cl_n

Polychlorinated dibenzofuran (PCDF)

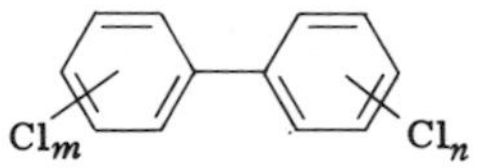

Cl_m Cl_n

Polychlorinated biphenyl (PCB)

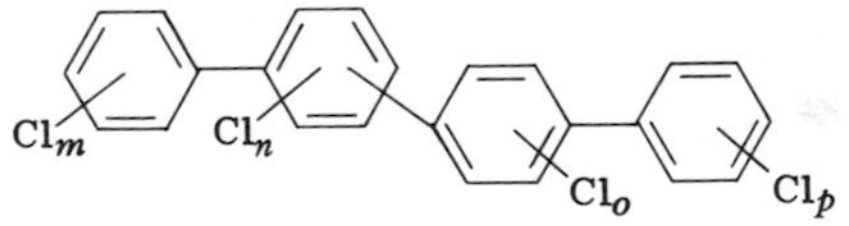

Cl_m Cl_n Cl_o Cl_p

Polychlorinated quaterphenyl (PCQ)

There are 209 possible isomers for the PCB's and 135 for the PCDF's. As can be seen in Table II, the concentration of these chemicals in the rice bran oil differed for Japan and for Taiwan. The toxicity of the oil from Japan was 25 times higher than that from Taiwan. Another difference was the distribution of the isomers. In Taiwan there was a higher concentration of the tetra- and pentachlorobiphenyls whereas in Japan, it was the hexa- and heptachlorobiphenyls. As for the PCDF's, the toxic oil contained isomers with 5 chlorine atoms

TABLE II. Concentration of PCBs, PCDFs and PCQs in Contaminated Oil in Japan and in Taiwan

	PCBs (ppm)	*PCDFs (ppm)*	*PCQs (ppm)*
Japan	920	5	866
Taiwan	40.5	0.26	36.5

TABLE III. Isomers of Toxic Oil in Japan and Taiwan

	Isomer	*Japan*	*Taiwan*
PCB's	Tetrachlor	-	+++
	Pentachlor	+	++++
	Hexachlor	++++	+
	Heptachlor	++++	-
PCDF's	Trichlor	+	+
	Tetrachlor-	++	++
	Pentachlor-	+++	++++
	Hexachlor-	++	++

with smaller amounts of the other congeners (Table III). This was true of the oil from both Japan and Taiwan.

The quantity of the ingested chemicals was calculated by epidemiological surveys (17,19,26), carried out in Japan in 325 cases and in Taiwan (924 cases)(Table IV). The surveys showed that there was a marked difference between the average consumption of rice oil per month in Japan (0.25 kg) as compared to Taiwan (1.42 kg). The ratio of contaminating chemicals ingested per month in Japan as compared to Taiwan was 1:4, respectively. The latency period to the onset of the first clinical symptoms was shorter in Japan (2 months) than in Taiwan (2.7 months). The total length of intake was also shorter in Japan (3.5 months) than in Taiwan (8.7 months). This is probably due to the fact that in Japan only a limited quantity of the rice oil was contaminated and only for a few days at the beginning of February, 1968. As soon as this

TABLE IV. Quantity of Ingested Toxic Oil and Duration of Ingestion[a]

Chemicals	*Quantity in rice oil (ppm=mg/kg)*		*Quantity rice oil used per month (mg)*		*Ingested during latency period (mg)*		*Ingested for total length of intake (mg)*		*Dose per day (mg)*		*Dose per kg body weight*	
	Japan	*Taiwan*	*Japan (0.25 kg)*	*Taiwan (1.42 kg)*	*Japan (2 months)*	*Taiwan (2.7 months)*	*Japan (3.5 months)*	*Taiwan (8.7 months)*	*Japan*	*Taiwan*	*Japan*	*Taiwan*
PCBs	920	40.5	230	57.5	460	155	805	500	8.4	1.9	0.14 mg	0.03 mg
PCDFs	5	0.26	1.25	0.37	2.5	0.99	4.4	3.2	0.04	0.01	0.06 µg	0.2 µg
PCQs	866	36.5	216.5	57.8	433	140	756	450	7.2	1.7	0.12 mg	0.02 mg

[a] *Values are calculated for average body weight of 60 kg. 325 cases in Japan, 924 cases in Taiwan.*

quantity had been consumed there was no additional opportunity for poisoning to occur. In Taiwan, it is not known how much of the rice oil had been contaminated and how long it had been available to consumers. The association of the clinical manifestations with the oil contamination was officially announced in October, 1979, i.e., 8 months after the observation of the first cases. The ratio in the total amount of PCB's, PCDF's, and PCQ's ingested in Japan (875 mg, 4.4 mg, and 756 mg, respectively) and Taiwan (500 mg, 3.2 mg, and 450 mg, respectively) is 1:1.6. These last values are probably more instructive with respect to the clinical manifestations observed in both places than the daily dose and the dose per kilogram body weight, where the ratio is roughly 1:4. The route of absorption for all cases was the gastrointestinal tract.

In Japan it was possible to estimate (17) the incidence (rate) of the clinical manifestations in the population in relation to dosage (Table V). It was shown that the toxic dose 100 (eliciting toxic effects in 100% of the cases) could be reached at a dose of 720-1400 mg PCB's, i.e., 3.6-7.2 mg PCDFs and 570-1100 mg PCQ's, and that below that level the incidence of disease decreased.

The clinical severity of the symptoms (27) graded by the dermatologist from 0 to 4 (Grade 0: subjective complaints; Grade 1: pigmentation of skin and mucosae, eye discharge; Grade 2: comedone formation; Grade 3: moderate acne: Grade 4: extensive acne eruptions) correlated closely in Japan with the total amount of oil consumed but not with the amount of oil consumed per kilogram of body weight per day. The acne lesions in the children were generally milder than those of the adults. The daily amount per kilogram body weight calculated for the children was larger than for the adults but the total amount was less. In Taiwan about ¼ of the cases were severe and clinical severity seemed to be higher in the younger age brackets. There seems to be a direct correlation between total intake of toxic oil and severity of symptoms, but this cannot be confirmed statistically (28-30).

VII. CLINICAL SYMPTOMATOLOGY

Probably the best (31) clinical picture was provided by the examination of a large Taiwanese family of 27 members, comprising three generations. The living conditions were the same and the toxic oil consumption had been identical for all members with respect to daily and total intake (Table VI). Table VI shows the clinical symptoms and signs in order of frequency and time of appearance. Symptoms affecting the eyes were the first to appear after a latency period of 2-3 months (32).

TABLE V. Frequency and Severity of Symptoms in Relation to PCB Concentration in Yusho Oil Japan 1968-1979[a]

Total number of users: 146	*Not affected (%)*	*Moderately affected (%)*	*Severely affected (%)*	*Attack rate (%)*
Light users (80)				
i.e., <720 ml or	10	39	31	87.5
less than 662 mg PCB's;	12	49	38.5	
623 PCQ's; 3.6 mg				
PCDF's				
Moderate users (45)				
i.e., 720-1440 ml				
or 662-1324 mg PCB's;		14	31	100
623-1247 mg PCQ's;				
3.6-7.2 mg PCDF's		31	69	
Heavy users (21)			21	100
i.e., more than			100	
1440 ml				

[a]*Oil contaminated with 920 ppm PCB; 866 ppm PCQ, and 5 ppm PCDF. After Kuratsune et al., 1972 (17).*

TABLE VI. Symptoms and Signs of 27 Members of a Three Generations Family 6-7 Months after Beginning of Poisoning, Taiwan, 1979[a]

Symptoms and signs	*Male (15)*	*Female (12)*
Increased eye discharge	93.5%	91.6%
Swelling of eyelids	86.6	91.6
Acneiform eruptions	86.6	83.3
Pigmentation of nails	86.6	83.3
Pigmentation of conjunctivae	80	83.3
Pigmentation of lips	80	66.6
Black color of nose	66.6	75
Hypoestesia	66.6	75
Deformity of nails	60	58.3
Pigmentation of gingivae	53.3	66.6
Numbness of limbs	53.3	50
Blurred vision	53.3	50
Keratotic plaques of palms and soles	46.6	66.6
Follicular hyperkeratosis	40	41.6

[a]*After Wei-Min Li et al., 1981 (31).*

The same clinical symptoms were observed in groups of cases examined and treated by the Department of Dermatology of the Veteran General Hospital, Taipei, Taiwan (33), as well as by the Departments of Ophthalmology, Neurology, Pediatrics, and Biochemistry of the same Hospital. A comparison of the dermatologic symptoms of two similar groups of cases from Japan shows that there are quantitative differences, which are probably related to differences in the constitutions of the samples, but the quality of the clinical features was the same (Table VII).

Comprehensive neurological examinations have been (34) performed on 39 cases of a group of 122 severe cases which were admitted to the Department of Dermatology of the Veteran General Hospital of Taipei between January and November, 1980.

TABLE VII. Comparison of Dermatologic Signs between 122 Cases in Taiwan (A) and 138 cases in Japan (B)

Sign	*Male*		*Female*	
	A *(42)*	*B* *(72)*	*A* *(80)*	*B* *(66)*
Acneiform skin eruptions (Comedones, follicular hyper-keratosis, yellow cysts, abscesses)	71.4%	86.1%	77.5%	77.3%
Dry Skin	50.0	30.6	56.0	36.4
Deformities of nails	68.5	22.2	67.5	24.2
Nail pigmentation	56.7	75.0	61.5	71.2
Black color of nose	48.7	4.2	57.3	4.5
Hyperkeratosis of palms and soles	19.5	1.4	22.5	4.5

TABLE VIII. Neurological Signs and Symptoms of 39 Cases Selected in a Group of 122 with Skin Lesions[a]

Symptoms and signs	*Number*
Headache	15 (38.4%)
Dizziness	12 (30.8%)
Paraestesia, numbness in the distal part of the limbs	25 (64.1%)
Hypoaestesia, hypoalgesia in the distal part of the limbs	13 (33.3%)
Pain over back, limbs, and orbits	16 (41.0%)
Intermittent blurred vision	11 (33.3%)

[a]*All reflexes were normal--no pathological reflexes; no muscle weakness or wasting.*

All had been treated unsuccessfully by local district doctors for 6-10 months. All cases had skin lesions of varying degrees of severity. The 39 cases referred to the Department of Neurology had neurologic symptoms upon clinical examination (Table VIII). The cases presented with a mild peripheral neuropathy with involvement of primary sensory neurons without inclusion of the motor neurons. There was also a mild involvement of the central nervous system. The electrographic measurement of nerve conduction velocity provided evidence that in 35 of the 39 cases not only sensory but also motor conduction velocity was slightly reduced. Nerves conduct at velocities ranging from 1 to 100 m/sec. The speed of nerve conduction indicates the functional status of a group of axons. The velocity in the PCB-contaminated cases was slightly longer than in a control group, averaging 63.3 m/sec. for the motor neurons and 52.9 m/sec for the sensory neurons, instead of 58.3 and 46.8 m/sec, respectively.

The neurological signs and symptoms also indicate involvement of the central nervous system, although electroencephalographic analysis did not show abnormalities.

In Japan, many Yusho patients complained of neurological symptoms, such as, numbness of limbs, hypo- and hyperesthesia, feeling of weakness, and muscle spasm. Reduced sensory nerve conduction velocity was observed in 9 out of 23 cases examined soon after poisoning, while reduced motor nerve conduction velocity was seen in only two cases.

Ultrastructural examination of the sciatic nerve of rats treated with PCB's did not show any change. No abnormalities of the central and peripheral nervous system have been produced in monkeys treated with 0.5 mg PCB's and 2.5 μg PCDF's for 6 weeks (35).

Blood chemistry tests were performed (36, 37) on several of the cases admitted to the Department of Dermatology of Taipei. The level of triglycerides was elevated in all age brackets. The SGOT and SGPT values, as well as those of the alkaline phosphatase, were within or at the upper borderline of the standard values, with the exception of the lowest age brackets (3 months of age). The blood picture and hemoglobin were normal.

Conventional function tests of the liver (bilirubin, SGOT, SGPT, BSP retention, and others) did not show any abnormalities of this organ in the Yusho cases. Although few morphologic changes were detected by light microscopy, the serum triglyceride was abnormal (38). The increase persisted for a few years (Table IX).

In about 20% of the Yusho cases chronic bronchitis was reported 10 years after the poisoning. The PCB concentration in the sputum and blood of these cases seemed to correlate with the severity of the symptoms. This has not thus far been reported in Taiwan.

VIII. PREGNANCY AND CHILD DEVELOPMENT

The course of pregnancy and subsequent child development was studied in 5 pregnant women (39,40) from Taiwan, who had conceived at the time they were using PCB-contaminated oil and who developed the most common dermatological manifestations (mucocutaneous pigmentation, acneiform eruptions, deformity of the nails, ocular manifestations, etc.). Birth was full term in 4 women and 1 month premature in one who gave birth to twins (Table IX).

Of the skin lesions, different degrees of generalized pigmentation, and deformity of the nails was observed in all of the babies. Although bone development was normal during the first year of life, the babies suffered from gastrointestinal and respiratory illnesses. They were not breast fed. Skin lesions and pigmentation disappeared 2-5 months after birth in all but one of the babies (case No. 4), who at the end of the first year of life still had no teeth and was unable to sit and to crawl. Retarded teeth and bone development was also observed in Japanese babies born from mothers who had been poisoned with the Yusho oil (41).

A comparison with findings in Japan has shown a similar general picture. Eleven women with overt symptoms of poisoning and two women married to men who had been exposed to the

TABLE IX. Clinical Findings of Newborn Babies of PCB-Contaminated Mothers, Taiwan, 1979

	Case number					
	1	*2*	*3*	*4*	*5*	*6*
Sex	Male	Male	Female	Male	Female twins	
Birth weight (gm)	3600	2500	2400	3600	1800	1800
Pigmentation of skin, nails, and mucosae	+	-	+++	++	++	++
Acneiform eruptions	+	-	+	++	-	-
Dry skin desquamation	+	+	+	++	+	+
Swelling of Meibomean glands	++	-	+++	+++	+	+
Deformity of nails	+ Both 5th fingers	-	+Both 5th fingers	+++	+	+
Black color of nose	+	-	+	+	-	-
Neurologic disorders	-	-	-	-	-	-

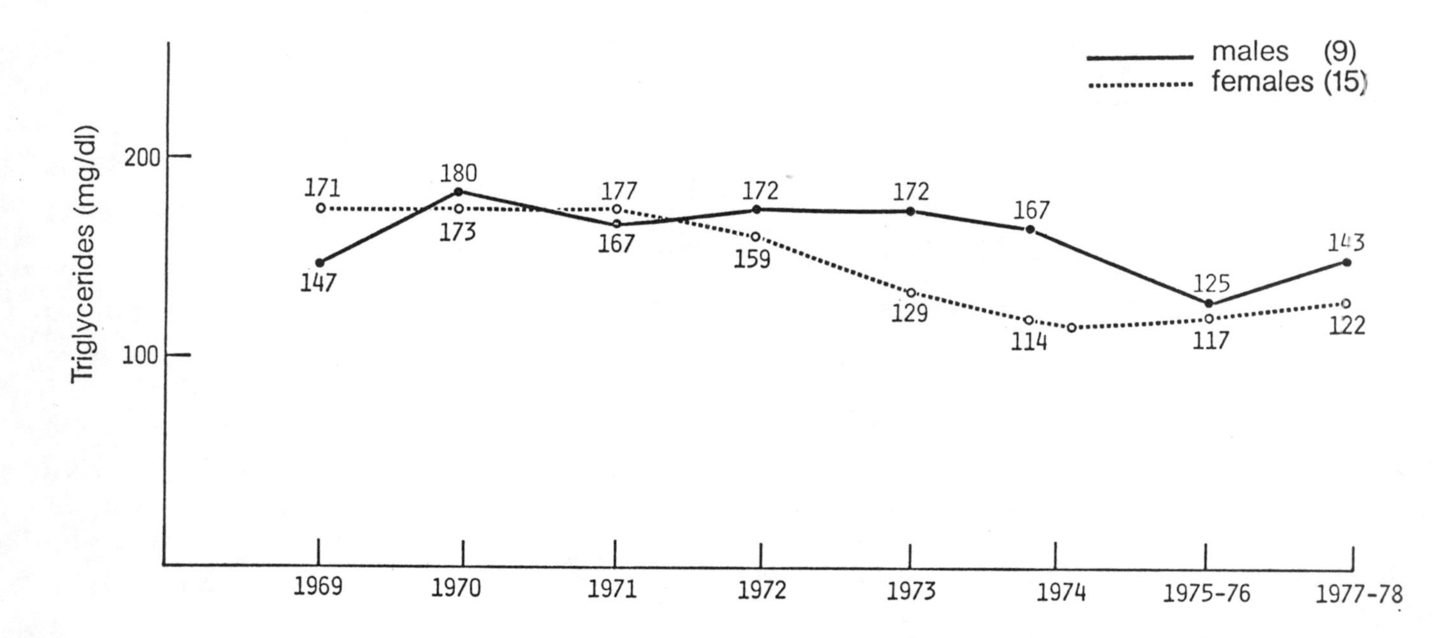

Fig. 1. Ten year follow-up of serum triglyceride levels in 24 Yusho patients (PCB+PCDF) Standard values: 47-162 mg/dl.

poison but who did not present overt symptoms of the disease delivered 11 live borns and two stillborn babies. Of these, 10 had dark brown pigmentation of skin mucosae and nails, as well as increased eye discharge. Five were considered to be undersized. A battery of laboratory tests, however, showed no abnormalities. The skin coloration faded within a few months.

Another woman suffering from the disease delivered 3 babies within 4 years after the incident occurred. All showed dark brown skin pigmentation. Pigmentation of the skin also occurred in a baby breast fed by a woman affected by the disease. However, no cases of unusual malformations or increased incidence of spontaneous abortions were reported.

IX. METABOLISM

The rate of absorption, metabolism, distribution, and elimination of these chemicals in humans are practically unknown. However, some data concerning blood levels, storage in adipose tissue and liver, and dissipation from these stores are available (42-44).

The concentration of these chemicals in the blood, liver, and adipose tissue was always higher in cases of poisoning than in the normal population (Table X). Blood PCB levels in normal subjects were below 4 ppb in Taiwan and were about 3 ppb in Japan. Levels in poison cases from Taiwan were ten times higher than normal; they are even higher than the levels found in Japanese cases who showed an average of 6.7 ppb. However the interpretation of these results are difficult because measurement of the blood PCB levels in the Japanese cases was performed 5 years (1973-1974) after the poisoning, whereas the same tests were carried out in Taiwan 1 year after the poisoning.

X. BLOOD AND TISSUE LEVELS AND RELATIONSHIP OF CLINICAL EFFECTS

The correlation between blood levels, severity of clinical symptoms, and blood chemistries has been examined many times, but no close relationship has been found to exist. A high PCB level in blood does not necessarily parallel severity of skin, lesions or rise in the serum triglyceride levels (45,46). On the other hand, blood and tissue concentrations of the chemicals seem to be higher and remain at that level in exposed as compared to the normal population (Table XI).

TABLE X. PCB Levels in Blood of 66 Cases of Contamination in Taiwan, 1979

Number	*Sex*	*Age*	*PCB levels (ppb)*
26	Male	31.4 (8-60)	47.5 (19-120)
37	Female	25 (12-69)	33.7 (11-71)
2	Female	2 months	18 (15-21)
37	Normal persons (Japan)		2.9 ± 1.7
	Normal persons (Taiwan)		Less than 4

TABLE XI. Concentration of PCB in Tissues of Yusho Patients[a]

		Concentration (ppm)	
Case no.	*Time of death or operation*	*Adipose tissue*	*Liver*
1	October, 1969	0.1	0.07
2	July, 1969	3.7	0.14
3	July, 1969	15	0.2
4	November, 1969	8.4	0.7
5	December, 1970	0.9	0.07
6	May, 1972	6.5	0.8
7	September, 1972	2.9	--
8	April, 1977	0.4	0.07
9	March, 1977	1.4	0.06
10	September, 1977	1.1	0.02
Control cases (31)	1973	>1.0	0.05

[a]*After Kikuchi et al., 1979 (46).*

Chemical dissipation time seems to be longer for some isomers than for others (Table XII). However, the interpretation of these findings is hampered by the limited number of the cases providing this data. Pentachloro isomers apparently have longer dissipation times than tetrachloro isomers. The distribution of the PCBs isomers also differs for cases of poisoning than for the normals (Table XIII). In the normal population there is a greater concentration of pentachloro than hexachloro isomers, whereas in the Japanese cases the hexachloro and heptachloro isomers are greater. This finding corresponds to the composition of the ingested oil.

XI. CONCLUSIONS

The ingestion of PCBs, PCDFs, and PCQs during several months, at an average daily dose of 1.9-8.4 mg PCBs, 0.01-0.04 mg PCDFs, 1.7-7.2 mg PCQ's, and a total dose of 500-805 mg PCB's, 3.2-4.4 mg PCDF's, and 450-756 mg PCQs resulted in a specific clinical picture, namely, skin manifestations dominated by acneiform lesions, pigmentation of skin, mucosae, and nails, and ophthalmic lesions, pigmentation of skin, mucosae, and nails, and ophthalmic lesions. In 20-30% of these cases, mild neurologic disorders of sensory and motor nerves were also observed. Conventional liver function tests as well as conventional blood analyses were normal. Serum triglycerides were threefold higher than standard values. Levels returned to normal after 3 to 5 years in the Japanese cases.

There are some general conclusions which can be drawn from the information gained by the two episodes which are possibly valid for all chemicals. The primary manifestation of the poisoning with these chemicals is skin lesions, which is also characteristic of those chlorinated chemicals which are chloracnegenic. Impairment of other physiologic functions, such as, the liver and nervous system, and fat and porphyrin metabolism are less frequent and less severe. It can be reasonably argued, further, that if the skin lesion does not appear other functions and organs are not affected. Of course, this does not hold if the contaminated chemicals are not chloracnegens (e.g., 2,4-DCP which is specifically neurotoxic and not chloracnegenic). Skin manifestations heal slowly (47) in about 80% of the cases. In some cases, sequelae remain which can be detected clinically when other symptoms and signs have disappeared. In addition, symptoms such as ocular lesions, peripheral neuropathies, liver function impairment, and increase of serum triglycerides return to normal from a few months to a few years following exposure, depending on the severity of the poisoning. As is frequently the case with all these chemicals, there is never a large increase in SGOT (SGPT indicates

TABLE XII. Concentration of PCDF Isomers in Tissues of Yusho Patients[a]

Case no.	Time of death	Tissue	2,3,6,8-Tetra-CDF	2,3,7,8-Tetra-CDF	1,2,4,7,8-Penta-CDF	2,3,4,7,8-Penta-CDF	Hexa-CDF	Total
2	July, 1969	Liver	0.7	0.3	7.1	6.9	2.6	17.6
3	July, 1969	Liver	0.08	0.02	0.4	1.2	0.3	2.0
		Adipose	0.6	0.3	1.0	5.7	1.7	9.3
6	May, 1972	Liver	0.03	0.01	0.09	0.3	0.03	0.46
		Adipose	0.08	n.d.	0.2	0.8	0.2	1.28
8	April, 1975	Adipose	0.4	n.d.	0.8	0.1	0.5	1.8
9	March, 1977	Liver	n.d.	n.d.	0.02	0.1	0.04	0.16
		Adipose	n.d.	n.d.	0.2	0.5	n.d.	0.7
10	Sept.,1977	Liver	n.d.	n.d.	n.d.	n.d.	n.d.	--
		Adipose	n.d.	n.d.	n.d.	n.d.	n.d.	--

[a] *Concentration in ppb. Detection limit 0.005-0.01 ppb.*

[b] *After Kuratsune et al., 1980 (17).*

TABLE XIII. Concentration of PCB Isomers in Blood of Yusho Patients and Normal Persons[a,b]

PCB isomers	*Yusho patients (N-9) (ppb)*	*Normal person (N=17) (ppb)*
2,4,5, 3', 4'-Pentachloro	0.13 ± 0.05	0.22 ± 0.09
2,3,4, 3',4'-Pentachloro	N.D.	0.05 ± 0.02
2,4,5, 2',4',5'-Hexachloro	0.52 ± 0.22	0.15 ± 0.07
2,3,4, 2',4',5'-Hexachloro	0.65 ± 0.21	0.17 ± 0.07
2,3,4, 5',3'4'-Hexachloro	0.28 ± 0.07	0.04 ± 0.02
2,3,4,5, 2',4', 5'-Heptachloro	0.44 ± 0.15	0.12 ± 0.07
2,3,4,5, 2',3', 4'-Heptachloro	0.28 ± 0.09	0.11 ± 0.04

[a] *Detection limit - 0.01 ppb. N.D. = less than 0.01 ppb.*

[b] *After Masuda, 1980.*

important hepatocellular lesions). There seems to be a correlation between severity of the clinical symptomatology and the amount of chemicals accumulated in the body.

Inferences about reproductivity are limited by the small number of cases reported. Conception seems to be possible even in cases of overt poisoning. Pregnancy seems to follow a normal course. Early fetal losses (abortions) have not been reported, but some late fetal losses (stillborn) may be related.

The newborn child is extensively exposed to the chemicals through the mother which leads to specific skin manifestations, to decreased birth weight, and retarded development, in some cases. Congenital anomalies have not been reported, with the exception of deranged tooth eruption and growth as a late manifestation.

The immune response of the newborns has not been examined. The rather high frequency of acute infections during the first year of life points to an impairment of immune capability.

In the Japanese episode the toxic oil contained PCB's with mostly 6 and 7 chlorine atoms isomers and only small amounts 5 chlorine isomers. More than 40 isomers of PCDF have been found in the Yusho oil (48), including as their major components the highly toxic 2,3,7,8-tetra- and 2,3,7,8-tetrachlorodibenzofurans. The ratio of PCB's to PCDF's is 200 for Yusho oil (Table XIV), while the ratio is 50,000 for kanechlor. The ratio in adipose tissue is practically the same as in the absorbed product, but it is five times higher in the liver.

The clinical features of the episodes in Japan and Taiwan are probably the result of a concurrent toxicity of the specific isomers of the mixture ingested. The dibenzofurans are probably mainly responsible for the chloracne lesions. Polychlorinated dibenzofurans have a strong chloracnegenic effect, whereas polychlorinated biphenyls are only weakly chloracnegenic.

It is not possible to assess the risk of the chemicals involved in these episodes. The LD^{50} for 2,3,7,8-TCDF varies from 5 to 10 μg/kg in the guinea pig to over 6000 μg/kg in the mouse, with rabbit and monkey in between (49). The total amount of PCDF's absorbed by the Yusho and Taiwan cases is 3-4 mg; the daily dose varied between 0.01 and 0.04 mg, i.e., 0.2-0.6 μg/kg.

Tetra and penta isomers of the PCDFs prevailed in the mixture but the actual amount of the 2,3,7,8-TCDF isomer is not known. It is most probably less than the dose which is lethal for the guinea pig.

The acute and chronic experimental toxicity of the mixture of chemicals involved in the Yusho and Taiwan episodes is not available. An extrapolation of some of the data produced by the study of some isomers has only limited value because a

TABLE XIV. Concentration of PCB's and PCDF's and Their Ratios in Yusho Cases[a]

	PCBs (ppm)	*PCDFs (ppm)*	*Ratio (ppm)*
Kanechlor (unused)	1,000,000	20	50,000
Yusho oil	~1000	5	200
Yusho patients			
consumption (mg)	875	4.4	198
adipose tissue (ppm)	1.3	0.009	
total (mg)	13	0.09	144
Liver			
concentration (ppm)	0.05	0.0013	
total organ (mg)	0.075	0.0019	39.5

[a] *After Masuda et al., 1980 (22).*

change in the number and position of the chlorine atoms in the molecule can modify the clinical picture. Therefore, not only the quantitative but also the qualitative prediction of the toxic effects in humans of PCDFs and PCBs is questionable, if the composition of the mixture varies. The impact of physical factors, i.e., heat on the technical products, will probably produce a constant change in ratio of the chemicals and of their isomers in the mixture and, therefore, their toxicity. The risks and health hazards of each case must be assessed on its own merits.

SUMMARY

Quantitative toxicology in animals has provided accurate and comprehensive data for some of the halogenated dioxins and related compounds. For many others it has only partially been carried out or is still lacking. In humans the information is lagging behind for several reasons. The determination of exposure should be improved, taking advantage of the increased sensitivity of analytical methods for assessing evidence of absorption in man. Clinical symptomatology should be estimated in relation to the kinetics of the chemicals in the body. Signs of physiological adjustments of organs and functions involved in the detoxification process should not be considered an adverse or irreversible health effect. Precise information about daily intake, length of exposure and total dosing, and latency to onset of effects and time of appearance of the symptoms and signs of the clinical picture of the poisoning are available only rarely. Human response to these chemicals is, at best, measured during episodes of heavy exposure.

The PCB's, PCDF's and PCQ's poisonings in Japan (Yusho) in 1968 and in Taiwan in 1979 offer opportunities to correlate clinical features to the dose absorbed and stored in the body. The follow-up studies of the Yusho cases also provide information about the course and evolution of the symptomatology.

The health effects observed in the two episodes can be indicative for the effects to be expected in cases of long-term exposure to the same chemicals at different dose levels.

REFERENCES

1. Kimbrough, R. D. (1980). Occupational exposure. "Halogenated Biphenyls, Terphenyls, Naphthalenes, Dibenzodioxins and Related Compounds". Elsevier North Holland, Amsterdam.
2. Cairns, T., et al. (1980). Review of the dioxin problem. Mass spectrometric analysis of TCDD in environmental samples. *Biomed. Mass Spectr.* 7, 11-12, 484-492.
3. Crow, K. (1981). The cutaneous and systematic signs of poisoning with chloracnegens in the human subject. *Symp. Chlorinated Dioxins and Related Compounds,* October 25-29, 1981, Arlington, Virginia.
4. Poland, A. and Glover, E. (1977). Chlorinated biphenyl induction of aryl hydrocarbon hydroxylase activity; a study of structure-activity relationship. *Mol. Pharmacol. 13,* 924-938.
5. Poland, A. and Glover, E. (1980). TCDD: studies on the mechanism of action. *Develop. Toxicol. Environ. Sci. 6,* 223-239.
6. Poland, A. et al. (1976). Stereospecific, high affinity binding of TCDD by hepatic cytosol. *J. Biol. Chem. 251,* 16, 4936-4946.
7. Poland, A. et al. (1979). Studies on the mechanism of action of chlorinated dibenzodioxins and related compounds. *Ann. N. Y. Acad. Sci. 320,* 214-230.
8. Baars, A. J. et al. (1980). Induction of rat hepatic glutathione-transferase activities by 2,3,7,8-TCDD. *Proc. Workshop Impact of Chlorinated Dioxins and Related Compounds on the Environment,* Oct. 22-24, 1980, Rome.
9. Leng, M. (1979). Comparative toxicology of various chlorinated dioxins as related to chemical structure. *CIPAC Proc. Symp. Ser. 1.*

10. Landrigan, P. J. et al. (1979). Cohort study of Michigan residents exposed to PBBs. Epidemiologic and immunologic findings. *Ann. N. Y. Acad. Sci. 320,* 284-294.
11. Billiant, L. B. et al. (1978). Breast milk monitoring to measure Michigan's contamination with PBB. *Lancet,* 643-646 (23 Sept.).
12. Caramaschi, F. et al. (1981). Chloracne following environmental contamination by TCDD in Seveso, Italy. *Intern. J. Epidemiol. 10,* 2, 135-143.
13. Kimbrough, R. D. (1981). Studies of human populations exposed to environmental chemicals. *Workshop Assessment of Multichemical Contamination,* April 28-30, 1981, Milan, Italy.
14. Harakuni, U. et al. (1979). Present state of Yusho patients. *Ann. N. Y. Acad. Sci. 320,* 273-283.
15. Kuratsune, M. (1980). Yusho. "Halogenated Biphenyls, Terphenyls, Naphthalenes, Dibenzodioxins and Related Compound." Elsevier North Holland, Amsterdam.
16. Kuratsune, M. et al. (1971). Yusho, a poisoning caused by rice oil contaminated with PCBs. *HSMHA Health Rep. 86,* 12, 1083-1091.
17. Kuratsune, M. et al. (1972). Epidemiologic study on Yusho, a poisoning caused by ingestion of rice oil contaminated with a commercial brand of polychlorinated biphenyls. *Environ. Health Perspect. 1,* 119-128.
18. Hsi-Sung Chen, P. (1981). Polychlorinated biphenyls. Environmental occurrence, biological and toxicological effects. *Clin. Med. (Taipei) 7,* 4-8.
19. Chung-Fu Lan et al. (1981). An epidemiological study on PCB poisoning in Taichung area. *Clin. Med. (Taipei) 7,* 96-100.
20. Hsi-Sung Chen, P. et al. (1981). Toxic compounds in the cooking oil which caused PCB poisoning in Taiwan. I. Levels of PCBs and PCDFs. *Clin. Mid. (Taipei) 7,* 71-76.
21. Hsi-Sung Chen, P. et al. (1981). Toxic compounds in the cooking oil which caused PCB poisoning in Taiwan. II. The presence of polychlorinated quaterphenyls and polychlorinated terphenyls. *Clin. Med. (Taipei) 7,* 77-82.
22. Masuda, Y. et al. (1980). Polychlorinated dibenzofurans and related compounds in patients with Yusho. *Proc. Workshop on Impact of Chlorinated Dioxins and Related Compounds on the Environment,* Oct. 22-24, 1980, Rome.
23. Kamps, L. R. et al. (1978). Polychlorinated quaterphenyls identified in rice oil associated with Japanese "Yusho" poisoning. *Bull. Environ. Cont. Toxicol. 20,* 589-591.
24. Yamaryo, T. et al. (1979). Formation of polychlorinated quaterphenyls by heating PCBs. *Fukuoka Med. Acta 70(4),* 88-92.

25. Masuda, Y. and Kuratsune, M. (1979). Toxic compounds in the rice oil which caused Yusho. *Fukuoka Med. Acta 70* (4), 229-237.
26. Hayabuchi, H. et al. (1979). Consumption of toxic rice oil by "Yusho" patients and its relation to the clinical response and latent period. *Food Cosmet. Toxicol. 17*, 455-461.
27. Po-Chak Cheng and King-Yin Liu (1981). Dermatopathological findings of PCB poisoning patients. *Clin. Med. (Taipei) 7*, 41-44.
28. Hsi-Sung Chen, P. (1981). Dermatological survey of 122 PCP poisoning patients in comparison with blood PCB levels. *Clin. Med. (Taipei) 7*, 15-22.
29. Asahi, M. et al. (1979). Dermatological symptoms of Yusho alterations in this decade. *Fukuoka Med. Acta 70*(4), 172-180.
30. Masakazu Asahi et al. (1981). Dermatological findings and their analysis in the general examination of Yusho in 1976-1980. *Fukuoka Med. Acta 72*(4), 223-229.
31. Wei-Min Li et al. (1981). PCB poisoning of 27 cases in three generations of a large family. *Clin. Med. (Taipei) 7*, 23-27.
32. Yao-An Fu (1981). Ocular manifestations of PCB poisoning and its relationships between blood levels and ocular findings. *Clin. Med. (Taipei) 7*, 28-34.
33. Wen-Jen Wang et al. (1981). Investigations on severe PCB poisoning In-patients. *Clin. Med. (Taipei) 7*, 62-65.
34. Lie-Gan Chia et al. (1981). Neurological manifestations in PCB poisoning. *Clin. Med. (Taipei) 7*, 45-61.
35. Yoshihara, S. et al. (1979). Preliminary studies on the experimental PCB poisoning in rhesus monkeys. *Fukuoka Med. Acta 70*, 4, 135-171.
36. Okumura, M. et al. (1979). Laboratory examination of the patients with PCB poisoning. *Fukuoka Med. Acta 70*, 4, 199-207.
37. Hirayama, C. (1979). Hepatocellular dysfunction in patients with PCB poisoning. *Fukuoka Med. Acta 70*, 4, 238-245.
38. Okumura, M. et al. (1979). Ten year follow-up study of serum triglycerides levels in 24 patients with PCB poisoning. *Fukuoka Med. Acta 70*, 4, 208-210.
39. Kit-Ching Wong and May-Yae Hwang (1981). Children born to PCB poisoning mothers. *Clin. Med. (Taipei) 7*, 83-87.
40. King-Lee Law et al. (1981). PCB poisoning in newborn twins. *Clin. Med. (Taipei) 7*, 88-91.
41. Miller, R. W. (1971). Cola-colored babies. Chlorobiphenyl poisoning in Japan. *Teratology 4*, 211-212.

42. Hsi-Sung Chen, P. et al. (1981). Levels and gas chromatographic patterns of PCB in the blood of intoxicated patients after ingestion of toxic cooking oil. *Clin. Med. (Taipei) 7,* 35-40.
43. Chen, P. H. et al. (1980). Levels and gas chromatographic patterns of polychlorinated biphenyls in the blood of patients after PCB poisoning in Taiwan. *Bull. Environ. Cont. Toxicol. 25,* 325-329.
44. Shigematsu, N. et al. (1979). Tissue distribution and biological effects of PCBs components, especially to the respiratory tract. *Fukuoka Med. Acta 70,* 4, 246-251.
45. Chiou-Jye Chen and Rey-Long Shen (1981). Blood PCB level and serum triglyceride in PCB poisoning. *Clin. Med. (Taipei) 7,* 66-70.
46. Kikuchi, M. et al. (1979). Autopsy report of two Yusho patients who died nine years after onset. *Fukuoka Med. Acta 70,* 4, 215-222.
47. Masakazu, A. et al. (1981). Dermatological findings and their analysis in the general examination of Yusho in 1976-1980. *Fukuoka Med. Acta 72,* 4, 223-229.
48. Rappe, C. et al. (1979). Identification of polychlorinated dibenzofurans retained in Yusho patients. *Chemosphere 4,* 259-266.
49. Moore, J. A. et al. (1979). Comparative toxicity of three halogenated dibenzofurans in guinea pig, monkey and mice. *Ann. N. Y. Acad. Sci. 320,* 151-163.

CHAPTER 3

RECLAMATION OF THE TCDD-CONTAMINATED SEVESO AREA

Luigi Noe

Ufficio Speciale per Seveso
Seveso (Milano), Italy

On July 10, 1976 a complex mixture of chemical products escaped from the safety valve of a vessel employed in batch production of trichlorophenol (TCP). These chemicals were deposited on an area south of the chemical plant of ICMESA, in the Municipality of Meda, bordering the town of Seveso, approximately 12 miles north of Milano.

As a consequence of this event, a large number of small animals namely, rabbits, chickens, and birds, died. After some weeks many children living in the same area developed chloracne.

After consultation with the Swiss chemists of the Zurich-based Givaudan Company, the technicians of the "Laboratorio Provinciale di Igiene e Profilassi" of Milano, at the request of the local Sanitary Authorities, determined that one of the main polluting compounds was dioxin (i.e., 2,3,7,8-tetrachlorodibenzo-*p*-dioxin). Using various sources and with the support of the Milan University (Institute of Farmacology and Farmacognosy) as well as of the Institute of Health (I.S.S.), an analytical method using gas chromatography coupled with a

ISBN 0-12-193160-9

mass-spectrometer was developed to determine the extent of the pollution. Evaluation of the polluted areas was undertaken within a relatively short time as a joint effort of all available personnel.

From the analytical results obtained, a map was drawn which indicated the three different "at risk" zones, designated A, B, and R (see maps in chapter by Pocchiari et al.).

The Lombardy Regional Institution which has, among its responsibilities, health control in the region, established an emergency plan which included the chemical monitoring of the environment and its inhabitants and the indemnification of local homeowners and industries which were affected by the contaminating cloud. The Italian Government intervened in this program by providing 40 billion lire (about 45 million dollars) in support and the Givaudan Company began paying damages.

The limits of the polluted areas were established mainly by taking into consideration the analytical sensitivity (0.01 μg in 1 kg of soil) and the limited toxicological information available in the literature (only on animals). The Regional Law No. 2 of January 17, 1977 fixed at 50 $\mu g/m^2$ the limit above which the area had to be barred to human inhabitants who, therefore, were compelled to leave their homes. Unfortunately, most were within the most polluted Zone A.

Police and army units were used to patrol the contaminated areas to prevent possible accidental contamination. In addition, a plastic fence, with iron net and poles, was built around the area.

In those areas where the pollution was below the 50 $\mu g/m^2$ but above 5 $\mu g/m^2$ school-age children and pregnant women were evacuated daily. Cultivation and animal breeding were prohibited as well as construction.

Finally, wherever the contamination was still detectable, but below the Zone B level (5$\mu g/m^2$), in a very large area including the territories of six adjacent municipalities (1430 ha), it was decided to prohibit cultivation and animal breeding. Construction was also limited.

Possible intervention to decontaminate the territory was also considered. In phase two of the decontamination plan one of the most efficient methods to deal with the polluted material was to consider incinerating the area at 1200°C in a rotary kiln, similar to that used in cement production, assuming, of course, that alternative methods did not guarantee the destruction of the dioxin. The capacity of the incineration kiln was 80-100 mt/day. Alternative decontamination methods, which had been previously approved by the scientific consultative committees, needed to be tested first to demonstrate their practicality. Only after positive results were attained by any alternative decontamination program could the incineration plan be withdrawn.

The rotatory kiln project was soon abandoned, however, for many reasons: (1) the danger that some dioxin might escape from the incinerator stack; (2) the fear that such a large reclamation plant might remain after the reclamation process was terminated and be used for other polluted materials (3) the time needed to construct and engineer a working pilot plant and the construction of an industrial-size plant would have required many years.

Many alternative decontamination methods were considered in the first months following the incident in order to devise other potentially simpler solutions. An advertisement, which appeared in the Italian newspapers, and a workshop organized by the Health Dept. of the Lombardy Region on September, 1976, stimulated a number of interesting ideas and projects, many of which were carefully evaluated by an "ad hoc" consultative committee. The Ufficio Speciale formed by the Lombardy Region specifically for the Seveso incident, acted as a screening unit for the several proposals received. Between the years 1977-1980, several methods of soil reclamation that had been proposed by Italian and foreign researchers were evaluated. Unfortunately, none could successfully be adapted from the laboratory-bench scale to the field, due to the many engineering, construction, operating, and safety problems posed by the fact that hundreds of thousands of metric tons of soil or, more generally, of waste had to be treated. In addition, the dioxin concentration in the soil varied greatly from a few micrograms per kilogram of soil to picograms per kilograms of soil level in the same highly polluted area (Zone A). Methods for the chemical, photochemical, and microbial degradation of TCDD are considered later in this volume.

The Special Office for Seveso is particularly interested in collecting information on these problems of degradation, since the results obtained in the clay and alluvial soil of Seveso of the reduction of TCDD concentration with time differed from that recorded in the US literature (Alvin L. Young et al. "Fate of TCDD in the Environment" October, 1976). In areas with a high concentration of TCDD the natural reduction was negligible, although based on percentage was appreciable in Zones B and R. This fact suggests that TCDD could be degraded by light and not by bacterial action.

Some alternative means of disposing waste from Seveso have also been evaluated: (1) incineration in the Atlantic Ocean; (2) direct dumping into the ocean; and (3) disposal in a salt pit. The international conventions on the use of the seas and, in the last case, the hostility of the population, made all these projects impracticable. Thus the most practical solution was to dispose of the polluted soil in basins, the peripheral parts of which were dampproofed with a layer of Bentonite (a mixture of gravel and sand), overlaid with one sheet of plastic material (high density polyethylene), and finally capped with clean soil. This solution was favorable in that it removed toxic waste from the area and prevented any dispersion in the surrounding territory. The procedure is similar to the one used for radioactive waste, for which several natural and artificial barriers are used to ensure that no single radioactive element may come into contact with the environment. In the case of Seveso a variety of barriers were available:

1. The dioxin is bound to the clay, which is rather abundant in the soil of the polluted zones. In addition, dioxin is not water soluble.

2. In the disposal of the polluted soil resulting from the reclamation work, the most polluted is deposited in the central part of the basin while the less polluted is deposited around the core along the protective sheeting.

3. The plastic sheet (high-density polyethylene, 2.5 mm thick) is welded so as to constitute a unique blanket.

4. A 15-cm thick foundation built with sand and Bentonite (reinforced concrete), has the appropriate characteristics of impermeability and plasticity.

In addition to all these barriers, the deposit cap is protected by a layer of "Gunite" (concrete reinforced with an iron net) to prevent possible damage from the outside. The entire deposit is covered with a 1-m thick layer of soil. A well with a drainage pump allows the extraction of water which gathers at the bottom of the deposit during the filling phase. The water is extracted with a pump and then analyzed. The barriers are controlled by inspecting the inferior part of the basin through the well. Periodical checks of the deposit integrity are performed by personnel of the Sanitary Engineering Dept. of the Technical High School of Milano to ensure that the soil disposal system remains efficient.

The disposal project began during the summer in 1980 and, at the beginning, was of an experimental nature. Later, however, the National Committee authorized that it be extended to the polluted zone north of Via Vignazzola, i.e., the most polluted area of the Zone A (approx. 10 ha). The entire subzone called Al, is now completely reclaimed. For the southern part of Via Vignazzola the Ufficio Speciale has suggested to the National Committee that the same system be used. For areas with less contamination the possibility is being considered of further reducing the dioxin density by diluting with uncontaminated soil.

Careful analysis precedes the establishment of a pollution map before reclamation begins. Analysis is also needed to establish when the soil scarification can be stopped according to the limits fixed by the National Committee.

The Regional Government required Givaudan Company to evacuate the ICMESA Plant. All operations were stopped immediately after the accident. In addition, the company was ordered to begin, at their expense, a decontamination project of B department, the one from which the toxic cloud originated. This work could only begin after the removal of the "incriminated" equipment. A number of possible options existed to solve the problem of decontamination of the chemical equipment of department B:

1. Construction of a giant monolith (a concrete casting) to enclose both the equipment and the building containing department B.

2. Dismantle and (a) construct a small monolith to enclose the equipment only or (b) chemically reclaim the dismantled equipment.

3. Dismantle the equipment using the so-called "nuclear method," and successively remove the highly contaminated material properly contained and packaged.

From May to July, 1981, the Givaudan Company charged the CNEN (National Committee for Nuclear Energy, based in Rome, a public institution emanating from the Ministry of Industry) to engineer the dismantling of the equipment and to dispose of the equipment using the "nuclear method". In addition, contact was to be avoided between the workers and dismantled equipment. The equipment was to be encapsulated in specially designed containers and was to be disposed in a remote location. It has not as yet been decided where this highly contaminated material could be deposited. A number of suggestions are available: disposal in a salt mine, burial in a controlled and protected site, and burial in an abandoned pit in a mountain region to be selected depending on seismic and geographic characteristics.

All workers operating in the polluted areas were given physical examinations to ascertain their fitness. Workers were required to wear protective clothing (complete suit, mask with filter, and gloves) and were thoroughly informed about the risks involved in working in the area.

Workers changed into protective clothing in a "Filter Station" at the borderline of the fenced Zone A. Using the same selection criteria a control group was also chosen; this group was checked every 6 months. By comparing the results of both medical examinations, no differences were found between the workers who were exposed to the toxic compound and the control group. A report illustrating the medical control methodology was presented in June, 1981 at the Epidemiology Congress of Helsinki (Finland).

This chapter emphasizes the technical aspects of the reclamation problem of the Seveso area, omitting other problems, such as political, sociological, epidemiological, which are beyond the scope presented here.

The search for a reclamation method has uncovered a number of organizational and operational problems, which the Italian Authorities are attempting to solve by a number of means. One is the establishment of a special authority under the responsibility of one Minister. Another is to elicit international cooperation in the regulatory and scientific fields. The European Community has prepared (but has approved) guidelines, the so-called "Direttiva Seveso," by which the European national regulations can be incorporated into laws and enforced, to prevent possible future environmental pollution by industry.

It is the author's opinion, that the best safeguard for our future is the rise of an ecological conscience by the entire industrial community.

DISCUSSION

DR. COULSTON: I am told that much of Professor Noe's paper was new information and it is of special interest on how they are meeting this problem in Italy.

DR. KOLBYE: In his paper, Professor Reggiani reported on the differences in the concentrations of PCB's and PCD's in the two oils in Japan and Taiwan. The Taiwan oil had the lower chlorinated isomers and, hence, we would expect greater acute toxicity. Your figures supported this very nicely in humans. The Yusho oil was contaminated with used Kanechlor. From the information we have, the used Kanechlor, during the reaction with the heat exchange process, was producing PCD's at approximately 500 ppm level in the Kanechlor.

On another issue, can anyone tell me about the number of offspring produced by the women in the Seveso area and whether the offspring were as normal as we have heard?

DR. COULSTON: This latter question will come up later and will be discussed in great detail. However, perhaps Dr. Bruzzi can give us a brief comment at this time.

DR. BRUZZI: I was not there at the time, but my group became involved later. Omitting many of the details, I can tell you that there were approximately 30 legal abortions. We do not, however, have information on many of them. In response to the question of whether or not the aborted fetuses were normal, no cases of abnormalities were observed, and all tissues examined were normal histologically.

ISBN 0-12-193160-9

DR. COULSTON: This issue will be discussed later in considerable detail.

DR. BLAIR: What was the official figure on the amount of dioxins liberated at the time of the Seveso accident? Can you tell us how you calculated the amount involved or perhaps it was a range?

DR. POCCHIARI: The amount of dioxins liberated in the Seveso accident is an important figure. Was there any agreed upon estimation? This is very difficult to determine, but the estimation ranges from 100 to 300 g. However, one cannot calculate the amount that disappeared into the atmosphere. The reactor contained 7000 kg. Before the process was stopped 2000 kg was liberated. The reactor went wild by excess heating; how long is not known. The temperature and the amount of time will determine the amount of dioxins formed. For example, the amounts in the pipeline, the roof of the building, and inside the plant can only be roughly estimated. By the calculations made, this was close to the several hundred grams estimated.

DR. COULSTON: An important issue in all this is the amount of TCDD retained and held in the soil.

DR. NIEMANN: In terms of human hazards the binding of TCDD to the soil is indeed important. This is an issue because the dioxins are tightly bound and, as such, pose little hazard. I know little about this except that microorganisms in the soil can help degrade dioxins.

DR. SILANO: Several years ago we investigated the possibility of using microorganisms to decontaminate the soil of Zone A. However, in Zone A the soil when so treated did not produce positive results and the work was stopped. Other workers have tried *Pseudomonas* in experimental decontamination, but the process was reported as impractical.

DR. COULSTON: Dr. Silano, you presented information that the level of dioxin in vegetation increased with the growing season from spring to summer and this also correlated with the airborne amounts of TCDD based on sample analysis. Can you separate out the amounts contributed from airborne as compared with the amounts from dioxin apparently held tenaciously in the soil?

DR. SILANO: Analyses were made of the soil in several different regions. We were especially interested in the toxicological significance of the 22 different possible isomers of TCDD as found in the ground and in the air in Zone A. Analyses made subsequently revealed that the distribution of the tetraisomers 1,3,6,8 and 1,3,7,9 varied. In

Zone R some other isomers were found that presumably came from other sources. However, 95% of the TCDD in Zone A was the 2,3,7,8 isomer. The levels of the other isomers are low and approach the limits of the methods used to detect them. Finally, it was found that of the numerous isomers present in Zone A, most frequent were the 1,3,6,8, 1,3,7,9, 1,2,3,4,6,7,8, and 1,2,3,4,6,7,9; others were likely present. We speculated, however, that some of these isomers may have come from municipal power plants. It is remarkable that the fly ash from combustion furnaces in different countries is now recognized to contain different ratios of these isomers. Free radicals formed during combustion may account for the differences in these complex mixtures. The entire subject is under investigation at this time, but it is strange that the ratios of the various isomers were the same in the vegetation and in fly ash, but there is no explanation for this fact.

OPENING REMARKS

DR. SHEPARD: Although I am not an expert on environmental toxicology, my role has been to oversee the Agent Orange program of the Veterans Administration as Special Assistant to the Chief Medical Director for Environmental Medicine. To the extent that I am not an environmental scientist I have been able to deal with the nonscientific issues. Many of the unknowns about the dioxins will be dealt with at this meeting. An International Symposium on Chlorinated Dioxins and Related Compounds has been organized for October 25-29, 1982 in Arlington, Virginia. These two meetings will complement each other.

ISBN 0-12-193160-9

CHAPTER 4

THE METABOLISM OF 2,3,7,8-TETRACHLORODIBENZO-*p*-DIOXIN IN MAMMALIAN SYSTEMS

J. R. Olson

Department of Pharmacology and Therapeutics
School of Medicine
State University of New York
Buffalo, New York

T. A. Gasiewicz

Environmental Health Sciences Center
University of Rochester School of Medicine
Rochester, New York

L. E. Geiger

Center in Toxicology
Vanderbilt University
Nashville, Tennessee

R. A. Neal

Chemical Industry Institute of Toxicology
Research Triangle Park, North Carolina

ISBN 0-12-193160-9

I. INTRODUCTION

Over the past decade, the public has become increasingly aware of health problems related to environmental and occupational exposure to chemical toxins. During this period, investigations in laboratory animals found 2,3,7,8-tetrachlorodibenzo-*p*-dioxin (TCDD) to be one of the most toxic synthetic chemicals known (1-4). As a result of its extreme toxicity and occasional presence in the environment, TCDD, commonly referred to as dioxin, has emerged as a compound for which there is great public health concern. A spectrum of toxic effects, similar to those observed with TCDD, can be produced upon exposure of animals to higher doses of other halogenated aromatic hydrocarbons. These include other polychlorinated dibenzodioxins, chlorinated dibenzofurans, certain polychlorinated biphenyl isomers and certain chlorinated azoxy- and azobenzenes (3, 5-7). TCDD is the most potent and one of the most extensively studied toxins in this group, and thus, has become a prototype for these ubiquitous, environmentally persistent, halogenated aromatic hydrocarbons.

TCDD can be formed as a low-level contaminant during the manufacture of 2,4,5,-trichlorophenol from 1,2,4,5-tetrachlorobenzene. It thus may occur as a contaminant of the herbicide 2,4,5-trichlorophenoxyacetic acid, and other products which use chlorophenols or chlorobenzenes as starting materials (3,4,8). A number of documented cases of accidental poisonings of animals and humans with TCDD have occurred over the years as a result of the production and use of these compounds (3,9-11). Perhaps the most widely publicized incident occurred in Seveso, Italy, in 1976, when an accident in a plant producing trichlorophenol resulted in the dissemination in excess of 100 gm of TCDD over a wide area that included several populated communities (11-13). The potential for widespread human exposure to chlorinated dioxins may exist with the recent discovery of these compounds in fly ash from municipal and industrial incinerators (14). It appears that various polychlorinated dibenzodioxins, including TCDD, may be formed as incomplete combustion products of organic chlorinated compounds (15). The potential for widespread exposure of man to low levels of TCDD has led to numerous studies attempting to establish man's relative sensitivity to the toxin. Studies to date, however, have been unable to estimate the acute or chronic toxicity of TCDD in man and have not established the mechanism(s) or site of action for the toxicity in experimental animals.

While investigations with laboratory animals have found TCDD to be one of the most potent toxic and teratogenic chemicals known (3,16), the mutagenic and carcinogenic activity of

Table I. Single Dose LD_{50} Values for TCDD

Species	*Route*	*LD_{50} (μg/kg)*	*References*
Guinea pig	Oral	2	*2*
Monkey	Oral	50	*45*
Rat			
Adult male	ip	60	*43*
Weanling male	ip	25	*43*
3-MC pretreated weanling male	ip	44	*43*
Adult female	ip	25	*43*
Rabbit	Oral	115	*1*
Rabbit	Skin	275	*1*
Mouse			
C57BL/6J	ip	132	*46*
DBA/2J	ip	620	*46*
$B6D2F_1/J^a$	ip	300	*46*
Hamster	ip	>3000	*22*
Hamster	Oral	1157	*22*

[a] *Offspring of C57BL/6J and DBA/2J, which are heterozygous at the Ah locus.*

the toxin are not as well established and depend on the experimental test system being used (16-21). The acute toxicity of TCDD in laboratory animals is characterized, in most cases, by listlessness, loss of body weight, delayed lethality (2 weeks to 2 months), and by marked interspecies variability in pathology and LD_{50}. Acute single dose LD_{50} values in various species have been summarized in Table I. Toxicity has been found to vary with the age, sex, and strain of the test animal, with the guinea pig being the most sensitive and the hamster the least sensitive species yet examined.

The extreme toxicity of some isomers of the polychlorinated dibenzodioxins has stimulated studies of their potential for widespread contamination of the environment and bioconcentration in certain species. TCDD, the prototype of this group of compounds, appears to be a remarkably stable compound, requiring temperatures in excess of 800°C for its thermal degradation (23). Once incorporated in the soil, TCDD appears to be extremely resistant to degradation by microorganisms, having a half-life of from 0.5 to > 10 years (24). On the other hand, the photolytic degradation of TCDD appears to be far more efficient, particularly if the toxin is in the presence of a suitable solvent which provides a hydrogen donor (24-26). Until recently, little was known about the fate of TCDD in biological systems. In 1979, Ramsey et al. (27) provided the first report suggesting that TCDD could be metabolized in mammals. This review will examine our current understanding of the metabolism of TCDD in mammalian systems. The significance of this data will also be discussed in relation to the toxicity of TCDD.

II. METABOLISM OF TCDD

A. *CHEMICAL NATURE OF TCDD-DERIVED RADIOACTIVITY IN TISSUES*

Analysis of the liver and adipose tissue of hamsters receiving ^{3}H- or ^{14}C-labeled TCDD indicated that >97% of the extractable radioactivity present was unmetabolized TCDD (28). These results confirm that of an earlier study which identified the radioactivity in the liver of ^{14}C-labeled TCDD-treated rats to be unchanged TCDD (29). Although additional studies are needed, these preliminary results suggest that TCDD remains largely in its unchanged form in those tissues of exposed hamsters and rats which have been examined.

The *in vivo* covalent binding of ^{3}H-labeled TCDD to rat liver protein, ribosomal RNA, and DNA has been reported (30). These investigators also found that virtually all of the radioactivity present in the liver (>99.9%) could be extracted. The distribution of the unextractable radioactivity was: protein, 60 pmol TCDD per mole of amino acid residue; RNA, 12 pmol TCDD per mole of nucleotide residue; and DNA, 6 pmol TCDD per mole of nucleotide residue. The maximum covalent binding of TCDD to DNA was calculated to be about one molecule TCDD per DNA, equivalent to 35 cells. This level of binding is 4 to 6 orders of magnitude lower than that of most chemical carcinogens.

B. *CHEMICAL NATURE OF IN VIVO EXCRETION PRODUCTS*

Some of the earlier studies with radiolabeled TCDD found no evidence that TCDD was metabolized either *in vivo* or *in vitro* (31-33). Studies on the elimination of ^{14}C-labeled TCDD in rats detected a small, unidentified portion of the administered radioactivity in the urine and expired air (29, 34,35). This finding provided evidence that some metabolic alteration or breakdown of TCDD may be occurring. In 1979, the excretion of unidentified metabolites of TCDD in rat bile was confirmed through chromatography, providing the first reports of the biological degradation of TCDD (27,36). Both investigations suggested that only metabolites of TCDD were eliminated in the bile of treated rats and that these metabolites may be glucuronide conjugates.

The chemical nature of the radioactive excretion products has been examined in the urine, feces, and bile of ^{3}H- and ^{14}C-labeled TCDD-treated hamsters (28,37). The radioactivity in the urine and bile was analyzed by high-performance liquid chromatography (hplc) and thin-layer chromatography (tlc) and was found to correspond to several polar metabolites of TCDD. Figures 1 and 2 show the hplc elution profiles of ^{14}C contained in urine and bile of hamsters administered ^{14}C-labeled TCDD. No radioactivity present in the urine or bile eluted from the column at the position of ^{14}C-labeled TCDD (fractions 46 and 47), suggesting that all of the radioactivity represented several metabolites of TCDD. The bottom panel of these figures show the hplc elution profiles obtained following the incubation of urine and bile with β-glucuronidase. In these chromatograms a greater percentage of the radioactivity was observed in the more nonpolar area (fractions 30-40) of the profile as compared to those samples which had not been treated with β-glucuronidase. No alterations of the elution profiles were observed upon chromatography of urine and bile which had been incubated with arylsulfatase. These results suggest that in the hamster, a large portion of the biotransformed TCDD was excreted in the urine and bile as glucuronide conjugates. In contrast to the TCDD-free excretion products of hamster urine and bile, extracts of fecal material from these animals were found to contain metabolites as well as unchanged TCDD (38). Unchanged TCDD accounted for about 25 to 45% of the radioactivity extracted from the feces of the hamster treated 4-9 days earlier with an ip dose of radiolabeled TCDD. Thus, it appears that significant quantity of unchanged TCDD may enter the intestinal lumen by a route other than the biliary pathway.

Radioactive urinary and biliary excretion products of ^{14}C-labeled TCDD-treated rats and guinea pigs have also been investigated (38). Figures 3 and 4 show the hplc elution patterns of ^{14}C contained in the urine and bile of ^{14}C-labeled TCDD-treated rats and guinea pigs, respectively. None of the radioactivity present in the urine and bile of these species eluted at the position of ^{14}C-labeled TCDD (fractions 46 and 47), suggesting that, as in the hamster, all of the radioactivity excreted in the urine and bile of rats and guinea pigs represents metabolites of TCDD. It is evident from

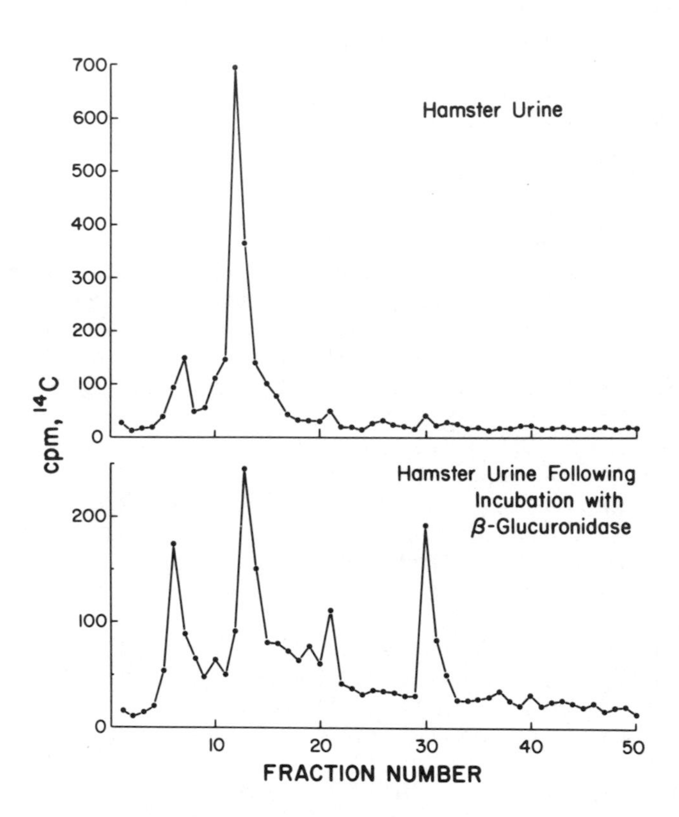

Fig. 1. High-performance liquid chromatographic separation of ^{14}C activity contained in urine from hamsters which were treated 7 days earlier with 500 µg ^{14}C-labeled TCDD/kg. The Altex Ultrasphere ODS (10 mm x 25 cm) column was eluted using a 30-min linear gradient of 99% water to 1% water in methanol at a flow rate of 3.0 ml/min. Fractions were collected at 1-min intervals and the radioactivity in each fraction was quantitated. Under these chromatographic conditions ^{14}C-labeled TCDD elutes in fractions 46 and 47 (37).

Figs. 1-4 that the elution times of the various radioactive metabolites in urine and bile are somewhat different in various species, suggesting some species differences in metabolism. The significance of these differences will have to await further work on the structural identification of these metabolites. The important point of these studies is that TCDD is converted to more polar metabolites, prior to elimina-

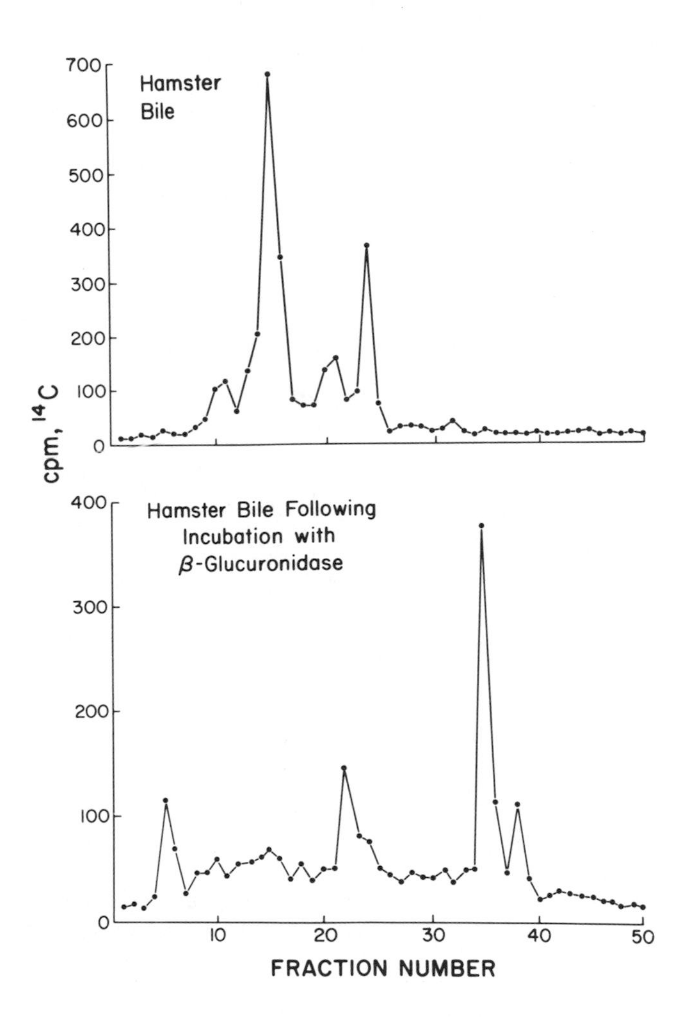

Fig. 2. High-performance liquid chromatographic separation of ^{14}C activity contained in bile from hamsters which were administered 7 days earlier 500 μg ^{14}C-labeled TCDD/kg. The chromatographic conditions are given in the legend to Fig. 1. Under these chromatographic conditions ^{14}C-labeled TCDD elutes in fractions 46 and 47 (37).

tion in the urine and bile of guinea pigs, rats, and hamsters. The apparent absence of these metabolites in extracts of hamster and rat liver (28,29) further suggests that once formed, the metabolites of TCDD are excreted in urine and bile.

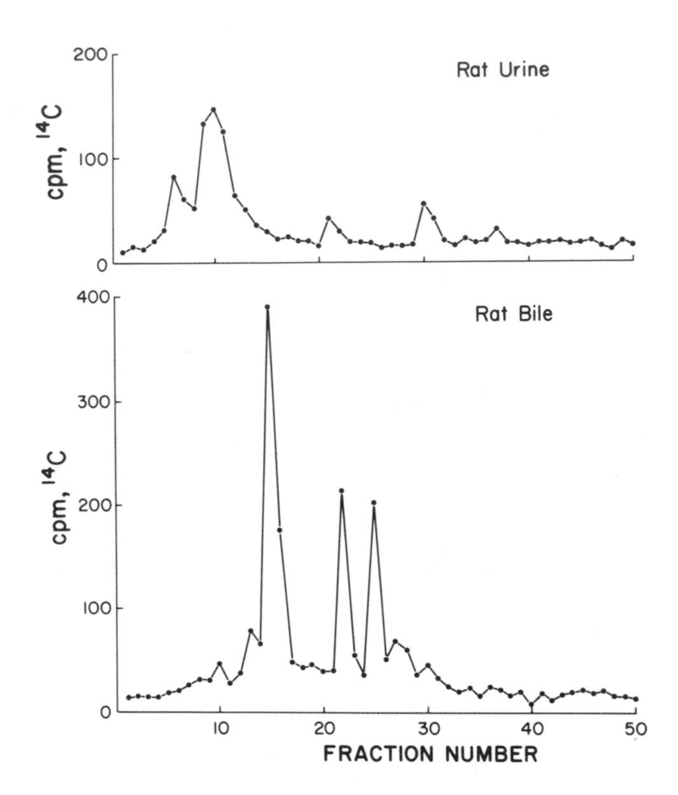

Fig. 3. High-performance liquid chromatographic separation of ^{14}C contained in the urine and bile of rats administered 7 days earlier 500 μg ^{14}C-labeled TCDD/kg. The chromatographic conditions are given in the legend to Fig. 1. Under these chromatographic conditions ^{14}C-labeled TCDD elutes in fractions 46 and 47.

C. METABOLISM OF TCDD BY HEPATIC MICROSOMES

TCDD metabolism in *in vitro* mammalian systems has also been examined. In one of the earliest studies, Vinopal and Casida (32) were unable to detect metabolism of TCDD by microsomal preparations from mouse, rat, and rabbit liver. They concluded that the chlorine atoms appear to greatly impede the metabolism of TCDD, since the nonhalogenated congener, dibenzo-*p*-dioxin, was rapidly converted to polar metabolites by liver microsomes (39). In other studies, unextractable radioactivity has been found to be bound to microsomes following the *in vitro* incubation of radiolabeled TCDD with rat (39) and mouse (40) hepatic microsomes. NADPH was found to be a

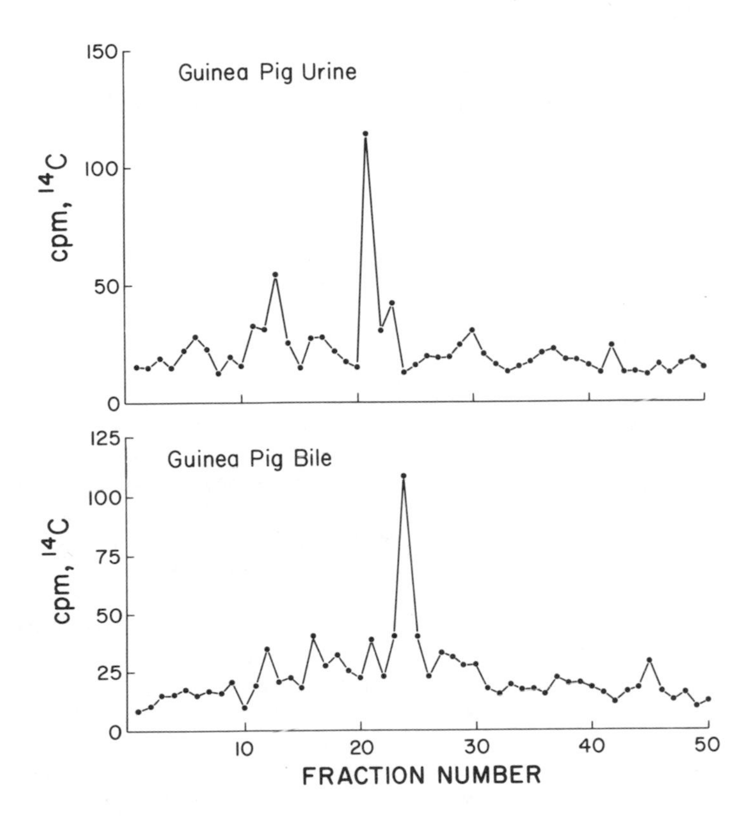

Fig. 4. High-performance liquid chromatographic separation of ^{14}C activity contained in the urine and bile of guinea pigs administered 7 days earlier 500 µg ^{14}C-labeled TCDD/kg. Chromatographic conditions are given in the legend to Fig. 1. Under these chromatographic conditions ^{14}C-labeled TCDD elutes in fractions 46 and 47.

necessary cofactor for the binding of radioactivity to the microsomal protein of mouse hepatic microsomes incubated with ^{3}H-labeled TCDD (40). In addition, carbon monoxide and α-naphthoflavone were found to decrease, while 3-methylcholanthrene pretreatment increased the irreversible binding of TCDD-derived radioactivity to microsomal protein. Although TCDD metabolites could not be extracted from the incubations, these authors presented the binding data as evidence for the metabolism of TCDD by the mouse liver cytochrome *P*-450-containing monooxygenase system (40). A recent study has examined the metabolism of highly purified ^{3}H-labeled TCDD by hamster liver microsomes (37). In these experiments, ^{3}H-labeled TCDD was purified by hplc immediately prior to the incubation. This procedure is essential, since in *in vitro* studies exceedingly small amounts of radioactive metabolites are detected. These experiments confirmed the earlier studies, demonstrating that NADPH was an essential cofactor needed for irreversible binding of TCDD-derived radioactivity to hepatic microsomal protein (37). Preliminary analysis of radioactive extracts from these incubations by hplc indicated the presence of metabolites of TCDD although the amount of radioactivity was too low to warrant additional experiments designed to identify the structure of these metabolites (37).

D. *METABOLISM OF TCDD BY ISOLATED HEPATOCYTES*

Isolated hepatocytes have been a valuable tool in the study of the biotransformation and conjugation of a large number of xenobiotics (41). Hepatocytes isolated from rats and hamsters have also been used to measure the metabolism of [1,6-^{3}H]TCDD (37,42). Using hamster hepatocytes, metabolites of TCDD could be readily detected (37). The hplc elution profile of radioactivity present in the cell-free media of incubations of ^{3}H-labeled TCDD with hamster hepatocytes is shown in Fig. 5. The data indicate the presence of at least four metabolites. The large peak at fractions 46 and 47 corresponds to unmetabolized ^{3}H-labeled TCDD. Radioactive peaks at fractions 10 and 24-27 correspond in retention time to metabolites found in hamster urine (Fig. 1) and bile (Fig. 2), respectively. The radioactive peak at fraction 5 in Fig. 5 may represent ^3H_2O. A similar peak has been found on chromatography of the urine and bile of ^{3}H-labeled TCDD-treated hamsters (38) and rats (36). The lower panel of Fig. 5 shows the elution profile of radioactivity from an aliquot of cell-free media which had been incubated with β-glucuronidase. As was the case with TCDD metabolites in hamster urine and bile (Fig. 1 and 2), some of the *in vitro* metabolites appear to be present as glucuronides. Control incubations using boiled hepatocytes or no hepatocytes were found to contain only ^{3}H-labeled TCDD. Thus, isolated hamster hepatocytes appear to be an effective *in vitro* model to examine TCDD metabolism. In addition, they have the ability to form TCDD metabolites with similar hplc elution times as those found in urine and bile.

E. *ROLE OF CYTOCHROME P-450 MONOOXYGENASE SYSTEMS IN TCDD METABOLISM*

Under the conditions of these experiments, isolated rat hepatocytes retained a greater than 90% viability for up to

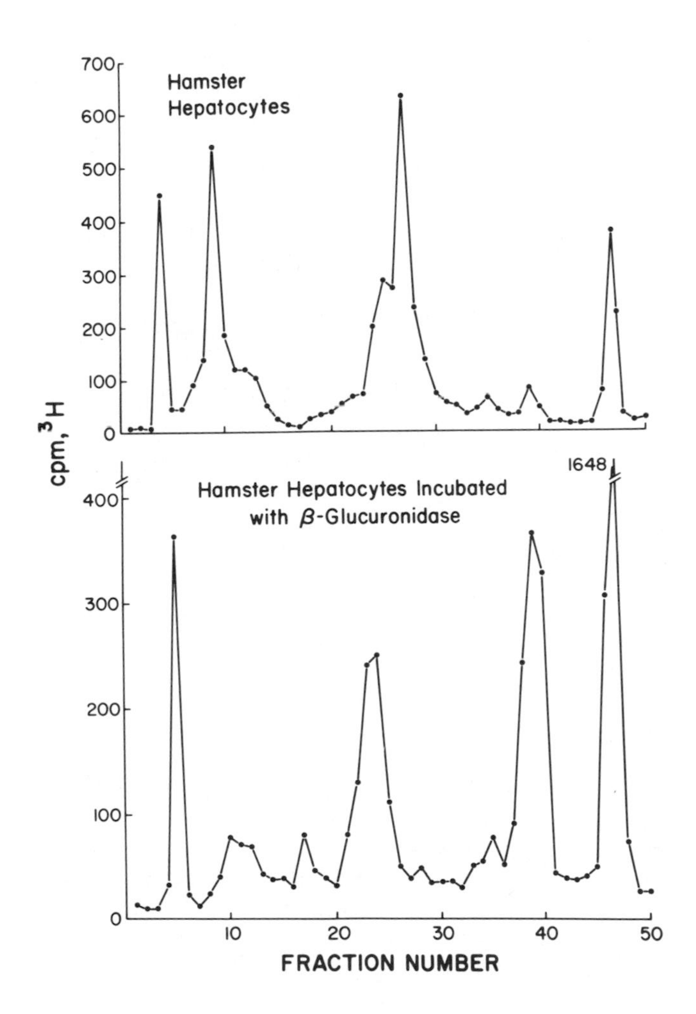

Fig. 5. High-performance liquid chromatographic separation of ^{3}H activity contained in the cell-free media of an incubation of isolated hamster hepatocytes with ^{3}H-labeled TCDD (2.0 μM). The chromatographic conditions are given in the legend to Fig. 1. Under these chromatographic conditions ^{3}H-labeled TCDD elutes in fractions 46 and 47 (37).

10 hours of incubation as assessed by trypan blue exclusion (42). Thus, they provided a good model to assess the rate of TCDD metabolism under various conditions (42). Rats were pretreated with the hepatic mixed-function oxidase (MFO) inducers, phenobarbital (PB) (80 mg/kg/day, ip x 3 days) or a single dose of TCDD (5 μg/kg, ip) or with the heme synthesis inhibi-

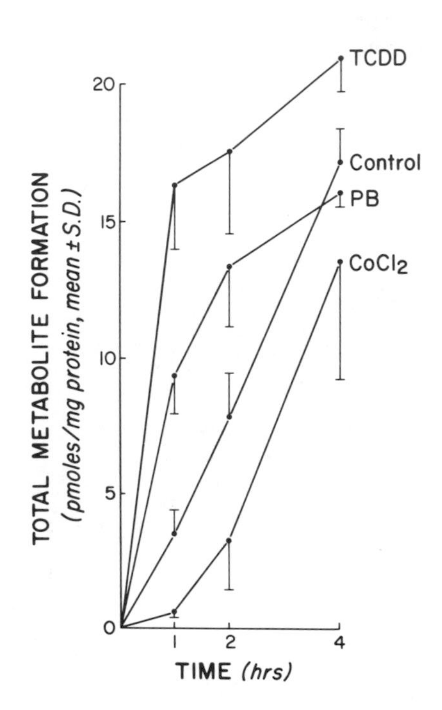

Fig. 6. The time course of TCDD metabolite formation in incubations containing ^{3}H-labeled TCDD (2.0 μM) and isolated rat hepatocytes. The cumulative rate of metabolite formation in untreated controls is compared with that of rats pretreated with phenobarbital (PB), TCDD, or $CoCl_2$. Each data point represents the mean ± SD of three experiments using hepatocytes isolated from different animals (42).

tor, $CoCl_2$ (60 mg/kg/day, sc x 2 days). TCDD and PB pretreatment was found to produce an increase in the rate of TCDD metabolite formation while $CoCl_2$ treatment reduced the rate of TCDD metabolism from that observed in the hepatocytes of untreated rats (Fig. 6). Thus, TCDD has the ability to induce its own metabolism. Hepatocytes from control and PB-pretreated rats were also incubated with ^{3}H-labeled TCDD in the presence of SKF 525-A (0.1 m*M*) or metyrapone (0.5 m*M*). These well-known inhibitors of drug metabolism reduced the rate of TCDD metabolism without producing any change in the viability of the hepatocytes (Fig. 7). Thus, the relative rate of TCDD metabolite formation in isolated rat hepatocytes appears to

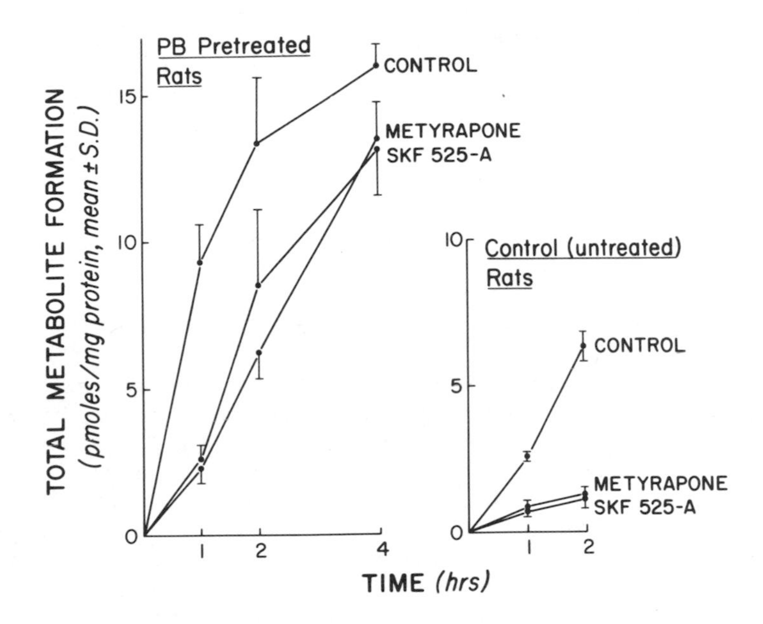

Fig. 7. Time course of TCDD metabolite formation during incubation of ^{3}H-labeled TCDD (2.0 μM) with isolated rat hepatocytes from control (untreated) and phenobarbital (PB) pretreated rats. The incubations were performed in the presence and absence (control) of SKF 525-A (0.1 mM) or metyrapone (0.5 mM). Each data point represents the mean ± SD of three experiments using hepatocytes isolated from different animals (42).

respond to drug-induced alterations in hepatic MFO activity. These results (37,42) and the results of others (40,43) suggest that TCDD is metabolized by the hepatic cytochrome *P*-450 monooxygenase systems.

F. *CHEMICAL STRUCTURE OF TCDD METABOLITES*

At present, the chemical structure of the TCDD metabolites remains unknown. It is possible that extrahepatic metabolism of TCDD is also taking place since metabolite profiles from urine have generally been found to be different from that of bile.

Incubation of metabolites from urine, bile, and hepatocyte incubations with β-glucuronidase results in products with longer hplc retention times (Fig. 1,2, and 5), implying that one or more metabolites may be glucuronide conjugates. The presence of glucuronides suggest the formation of phenolic derivatives of TCDD. It is possible that an arene oxide of TCDD is the initial intermediate formed. This product could then be converted enzymatically or nonenzymatically to dihydrodiol or phenol derivatives of TCDD. A number of reactive electrophilic arene oxide intermediates have been proposed to be responsible for the irreversible binding of TCDD-derived radioactivity to hepatic protein and other macromolecules (40).

G. ROLE OF METABOLISM IN THE TOXICITY OF TCDD

Although there have been no studies of the toxicity of TCDD metabolites, indirect evidence suggests that metabolism may be a detoxification process. Starting with the assumption that TCDD is metabolized by the hepatic mixed-function oxidase (MFO) enzyme system, one study examined the effects of altering the hepatic MFO activity on the toxicity (20-day LD_{50}) of TCDD in rats (43). These investigators used naturally occurring age- and sex-related differences in hepatic MFO activity in rats and observed that animals with lower MFO activity also had a lower 20-day LD_{50} for TCDD (see Table I). On the other hand, the LD_{50} values for TCDD in weanling rats were found to be increased by pretreatment with the MFO inducers, phenobarbital and 3-methylcholanthrene. The observed relationship between hepatic MFO activity and the LD_{50} for TCDD, led to the speculation that TCDD was metabolized by the hepatic MFO enzyme system in rats and that this metabolism resulted in the formation of a chemical species which was less toxic than the parent compound (43). This study thus suggested that metabolism was a detoxification process. Verification of this hypothesis must await the demonstration of increased excretion of TCDD metabolites in response to elevated hepatic MFO activity, and the direct assessment of the toxicity of these metabolites.

The apparently rapid elimination of TCDD metabolites upon formation suggests that the *in vivo* half-life for elimination of TCDD-derived radioactivity may provide a good estimate of the rate of TCDD metabolism in a given species. This data may offer additional information as to the relationship between the ability of an animal species to metabolize TCDD and the toxicity of TCDD in that species. Table II summarizes the acute LD_{50} and half-life for elimination ($t_{1/2}$) data for TCDD in various species and three strains of mice.

The guinea pig is the most sensitive species to acute TCDD toxicity. However, there was essentially no difference in the half-life of elimination values of 30 and 31 days for TCDD-

TABLE II. Rates of Elimination and Toxicity of TCDD in Various Species

Species	*Dose (μg/kg)*	*Half-life for elimination (days)*	*Relative percentage of TCDD-derived radioactivity*			
			Feces	*Urine*	*LD_{50} (μg/kg)*	*Reference for elimination*
Guinea pig	2.0 (ip)	30	94	6	1-2	*44*
Rat	1.0 (Oral)	31	>99	<1	60	*29*
Mouse						
C57BL/6J	10.0 (ip)	17	74	26	132	*46*
DBA/2J	10.0 (ip)	37	70	30	620	*46*
$B6D2F_1/J$[a]	10.0 (ip)	17	73	27	300	*46*
Hamster	650 (ip)	11	59	41	>3000	*28*
Hamster	650 (Oral)	15	--	--	1157	*28*

[a]*Offspring of C57BL/6J and DBA/2J, which are heterozygous at the Ah locus.*

derived radioactivity in the guinea pig (44) and the rat (29), respectively, suggesting that different elimination rates could not explain the species difference which was observed in toxicity.

The mean half-life for elimination of TCDD-derived radioactivity in the hamster was found to be approximately 11 and 15 days following the ip and po treatment with ^{3}H-labeled TCDD, respectively (28). This rate of elimination of TCDD-derived radioactivity is greater than that observed in the rat and guinea pig. This may contribute to, but does not appear to, totally explain the marked resistance of the hamster to TCDD toxicity.

The excretion of TCDD in C57BL/6J and DBA/2J mice has also been examined (46). These strains are considered to be responsive and nonresponsive, respectively, to aryl hydrocarbon hydroxylase (AHH) induction by 3-methylcholanthrene. The C57BL/6J strain was found to be more sensitive to the toxic effects of TCDD, having an LD_{50} of 132 μg/kg, which is approximately one-fifth the LD_{50} in the DBA/2J strain (Table I). In conjunction with the lower LD_{50}, the responsive C57BL/6J strain was found to have a half-life for elimination of TCDD of 17 days which is approximately one-half of that observed for the DBA/2J strain (Table II). The difference in elimination rate may in part be related to the finding that the DBA/2J mice possess approximately 60% greater adipose tissue stores than the C57BL/6J mice (46,47). Decad et al.(47) found a similar relative difference in the half-life of elimination values of ^{14}C-labeled 2,3,7,8-tetrachlorodibenzofuran in the C57BL/6J and DBA/2J strains. Thus, other factors, in addition to metabolic capability, may influence the rate of elimination of these compounds. In addition, Gasiewicz et al. (46) have also found that the $B6D2F_1$/6J mice (offspring of C57BL/6J and DBA/2J which are heterozygous at the *Ah* locus) have a half-life of elimination value identical to the C57BL/6J mice, in spite of having an intermediate LD_{50} value. The amount of adipose tissue stores in the C57BL/6J and $B6D2F_1$/J mice were also found to be similar (46). Thus, the relationship between toxicity and rate of elimination of TCDD in these strains of mice appears to be the opposite of that expected from earlier studies in the rat, guinea pig, and hamster, where the hamster eliminated TCDD at a significantly faster rate. These data suggest that no clear relationship exists between the ability of a given species to excrete TCDD and/or its metabolites and the toxicity of TCDD in that species as determined by the LD_{50} value.

III. CONCLUSIONS

Studies in rats and hamsters have indicated that the TCDD in these tissues was the parent compound rather than metabolites. In most species, TCDD-derived radioactivity was found to be largely eliminated in the feces. In the hamster, urinary elimination was also found to be a major route of excretion. All of the TCDD-derived radioactivity excreted in the urine and bile of hamsters, guinea pigs, and rats appeared to be metabolites of TCDD. The apparent absence of TCDD metabolites in liver and fat would suggest that once formed, the metabolites of TCDD were readily excreted. Although urine and bile appeared to be free of unmetabolized TCDD, data from the hamster indicated that a significant amount of unchanged TCDD may have been excreted into the intestinal lumen by some route other than bile for a number of days following treatment.

In contrast to hepatic microsomes, isolated hepatocytes are a good *in vitro* model for study of the metabolism of TCDD. Incubation of TCDD with hepatocytes produced metabolites of TCDD with similar hplc elution times as those found for metabolites of TCDD excreted in urine and bile. The relative rate of TCDD metabolite formation in isolated rat hepatocytes was found to be increased by prior administration of inducers of the hepatic mixed-function oxidase (MFO) enzymes. These and other studies suggest that TCDD is metabolized by the cytochrome *P*-450-containing monooxygenases.

At present the chemical structure of the TCDD metabolites is unknown. Incubation of metabolites found in urine, bile, and in hepatocyte incubations with β-glucuronidase indicates some of the metabolites may be glucuronide conjugates.

Although these studies suggest that metabolism of TCDD is a detoxification process, the exact role of the metabolism of TCDD in contributing to the varied toxicity of this compound in numerous animal species is unknown. At this time there appears to be no consistent relationship between the ability of a given species to excrete TCDD and/or its metabolites and the toxicity of TCDD in that species as determined by the acute LD_{50} values. However, the contribution of metabolism in the half-life for elimination values of TCDD must be interpreted with caution since other factors such as elimination of free TCDD in the feces and the amount of adipose tissue stores in the animal, in addition to metabolic capability, may influence the rate of elimination of these compounds. Furthermore, there may be additional factors other than metabolism contributing to a species relative sensitivity to the toxic effects of TCDD.

ACKNOWLEDGMENTS

This paper is based on work performed partially under Contract No. DE-AC02-76EV03490 with the U. S. Department of Energy at the University of Rochester Department of Radiation Biology and Biophysics and has been assigned Report No. UR-3490-2045.

REFERENCES

1. Schwetz, B. A., Norris, J. M., Sparschu, G. L., Rowe, V. K., Gehring, P. J., Emerson, J. L., and Gerbig, C. G. (1973). *Environ. Health Perspect.* *5,* 87.
2. McConnell, E. E., Moore, J. A., Haseman, J. K., and Harris, M. W. (1978). *Toxicol. Appl. Pharmacol.* *44,* 335.
3. Kimbrough, R. D. (1974). *CRC Crit. Rev. Toxicol.* *2,* 445.
4. IARC. (1977). *IARC Monographs on the Evaluation of the Carcinogenic Risk of Chemicals to Man, i.e., Some Fumigants, the Herbicides 2,4-D and 2,4,5-T Chlorinated Dibenzodioxins and Miscellaneous Industrial Chemicals, Lyon, France.*
5. Goldstein, J. A. (1979). *Ann. N. Y. Acad. Sci.* *320,* 164.
6. McConnell, E. E. (1980). *In* "Halogenated Biphenyls, Terphenyls, Naphthalenes, Dibenzodioxins and Related Products" (R. D. Kimbrough, ed.), p. 109. Elsevier/North Holland Biomedical Press, N.Y.
7. McConnell, E. E., and Moore, J. A. (1979). *Ann. N. Y. Acad. Sci.* *320,* 138.
8. Crummet, W. B., and Stehl, R. H. (1973). *Environ. Health Perspect.* *5,* 15.
9. Rawls, R. L., and O'Sullivan, D. A. (1976). *Chem. Eng. News,* August 23, 27.
10. Huff, J. E., Moore, J. A., Saracci, R., and Tomatis, L. (1980). *Environ. Health Perspect.* *36,* 221.
11. Firestone, D. (1978). *Ecol. Bull. (Stockholm),* *27,* 39.
12. Reggiani, G. (1978). *Arch. Toxicol.* *40,* 161.
13. Reggiani, G. (1980). *J. Toxicol. Environ. Health* *6,* 27.
14. Bumb, R. R., Crummett, W. B., Cutie, S. S., Gledhill, J. R., Hummel, R. H., Kagel, R. O., Lamparski, L. L., Luoma, E. V., Miller, D. L., Nestrick, T. J., Shadoff, L. A., Stehl, R. H., and Woods, J. S. (1980). *Science* *210,* 385.
15. Hay, A. (1981). *Nature (London)* *289,* 351.
16. Poland, A. and Kende, A. (1976). *Fed. Proc.* *35,* 2404.
17. Hussain, S., Ehrenberg, L., Lofroth, G., and Gejvall, T. (1972). *Ambio* *1,* 32.

18. Kociba, R. J., Keyes, D. G., Beyer, J. E., Carreon, R. M., Wade, C. E., Dittenber, D., Kalnins, R., Frauson, L., Park, C. N., Barnard, S. D., Hummel, R. A., and Humiston, C. G. (1978). *Toxicol. Appl. Pharmacol. 46,* 279.
19. Van Miller, J. P., Lalich, J. J., and Allen, J. R. (1977). *Chemosphere 6,* 537.
20. Walsh, J. (1977). *Science 197,* 1064.
21. Kouri, R. E., Rude, T. H., Joglekar, R., Dansette, P. M., Jerina, D. M., Atlas, S. A., Owens, I. S., and Nebert, D. W. (1978). *Cancer Res. 38,* 2777.
22. Olson, J. R., Holscher, M. A., and Neal, R. A. (1980). *Toxicol. Appl. Pharmacol. 55,* 67.
23. Stehl, R. H., Papenfuss, R. R., Bredeweg, R. A., and Roberts, R. W. (1973). *In* "Chlorodioxins-Origin and Fate" (E. Blair, ed.), p. 121. American Chemical Society, Washington, D. C.
24. di Domenico, A., Silano, V., Viviano, G., and Zapponi, G. (1980). *Ecotoxicol. Environ. Safety 4,* 339.
25. Crosby, D. G., and Wong, A. S. (1976). *Science 195,* 1337.
26. Nestrick, T. J., Lamparski, L. L., and Townsend, D. I. (1980). *Anal. Chem. 52,* 1865.
27. Ramsey, J. C., Hefner, J. G., Karbowski, R. J., Braun, W. H., and Gehring, P. J. (1979). *Toxicol. Appl. Pharmacol. 48,* A162.
28. Olson, J. R., Gasiewicz, T. A., and Neal, R. A. (1980). *Toxicol. Appl. Pharmacol. 56,* 78.
29. Rose, J. Q., Ramsey, J. C., Wentzler, T. H., Hummel, R. A., and Gehring, P. J. (1976). *Toxicol. Appl. Pharmacol. 36,* 209.
30. Poland, A. and Glover, E. (1979). *Cancer Res. 39,* 3341.
31. Fries, G. F., and Marrow, G. S. (1975). *J. Agr. Food Chem. 23,* 265.
32. Vinopal, J. H. and Casida, J. E. (1973). *Arch. Environ. Contam. Toxicol. 1,* 122.
33. Piper, W. N., Rose, J. Q., and Gehring, P. J. (1973). *In* "Chlorodioxins-Origin and Fate" (E. Blair, ed.), p. 85. American Chemical Society, Washington, D. C.
34. Piper, W. N., Rose, J. Q., and Gehring, P. J. (1973). *Environ. Health Perspect. 5,* 241 (1973).
35. Allen, J. R., Van Miller, J. P., and Norback, D. H. (1975). *Food Cosmet. Toxicol. 13,* 501.
36. Poiger, H., and Schlatter, Ch. (1979). *Nature (London) 281,* 706.
37. Neal, R. A., Olson, J. R., Gasiewicz, T. A., and Gudzinowicz, M. (1982). *In* "Halogenated Hydrocarbons: Health and Ecological Effects" (M. A. Q. Khan, ed.), in press. Pergamon Press, Inc., Elmsford, N.Y.

38. Olson, J. R. and Neal, R. A. (1980). Unpublished observations.
39. Nelson, J. O., Menzer, R. E., Kearney, P. C. and Plimmer, J. R. (1977). *Bull. Environ. Cont. Toxicol. 18*, 9.
40. Guenthner, T. M., Fysh, J. M., and Nebert, D. W. (1979). *Pharmacology 19*, 12.
41. Fry, J. R. and Bridges, J. W. (1977). *In* "Progress in Drug Metabolism" (J. W. Bridges and F. L. Chasseaud, eds.), Vol. 2, p. 71. Wiley, N.Y.
42. Olson, J. R., Gudzinowicz, M., and Neal, R. A. (1981). *Toxicologist 1*, 69.
43. Beatty, P. W., Vaughn, W. K., and Neal, R. A. (1978). *Toxicol. Appl. Pharmacol. 45*, 513.
44. Gasiewicz, T. A. and Neal, R. A. (1979). *Toxicol. Appl. Pharmacol. 51*, 329.
45. McConnell, E. E., Moore, J. A., and Dalgard, D. W. (1978). *Toxicol. Appl. Pharmacol. 43*, 175.
46. Gasiewicz, T. A., Geiger, L. E., and Neal, R. A. (1981). Unpublished observations.
47. Decad, G. M., Birnbaum, L. S., and Matthews, H. B. (1981). *Toxicologist 1*, 65.

DISCUSSION

DR. BARNES: Are there any other reasons for these specific differences found in these several species?

DR. NEAL: There is no difference in the receptors and their affinity for TCDD, and there is no correlation with the acute toxicity in this regard. How this substance causes its toxicity has not been determined. It seems to effect some fundamental aspect of the cell. In the guinea pig, if it were equally distributed, toxic doses are equal to 200 molecules per cell, which is an extremely small amount. There is no similar sensitivity in other species and, by comparison, the hamster exhibits remarkable resistance. Toxicity is simply not related to tissue levels. Unfortunately, there are no data on species differences in terms of subacute administration. Small, sublethal doses cause toxicity with long-lasting effects in some species, but systematic studies have not been made. There is no study yet published on the long-term effects of possible TCDD enhancement of nitrosamine toxicity in hamsters, although such work will be published in due time.

The important fact remains, which of the numerous animal species studies should be considered as closest to man? Although we do not know as yet, it appears that man more strongly resembles the less sensitive hamster than the most sensitive guinea pig, a conclusion which is based on the evidence obtained on man from substantial industrial exposures. If man were as sensitive as a guinea pig, the amounts known to be exposure doses would have been lethal and this has not been observed. It seems that with all these studies that have been made we should be able to select which animal surrogate most

ISBN 0-12-193160-9

closely approaches man with respect to toxicity of TCDD. However, until we know what the biochemical mechanisms of toxicity are, we can only say that man is probably as resistant as the hamster to the acute lethal effects of TCDD. At present, we cannot make any predictions beyond acute lethal effects, because we do not know what the true chronic effects in man are. Strain differences in mice and rats exhibit interesting effects that warrant additional study. As illustrated, mice are the species of choice to indicate strain differences.

DR. FRAWLEY: Have reproduction studies been done and is there any evidence of species differences based on subacute studies?

DR. NEAL: Estrogen effects may account for the sex differences; the female rat is less susceptible in some strains and more susceptible in others. This may be related to enzyme differences, since the castrated female more closely resembles the male.

Small doses of TCDD administered chronically will increase the LD_{50}. The animal then becomes more resistant to TCDD when a large acute dose is administered. This agrees with the finding that the liver hepatocytes are now better able to metabolize TCDD after administration of small doses of TCDD.

DR. KOLBYE: Are there any other long-term studies on hamsters?

DR. NEAL: Not that I know of.

DR. MRAK: Are there marked strain differences in rats?

DR. NEAL: Mice exhibit the greatest degree of strain differences. Estrogens may account for the sex differences.

DR. COULSTON: At some point, we must determine which animal species would most likely be selected to predict effects for man. Is there any choice at this point? What about the monkey?

DR. NEAL: As I have said, man is clearly more like the hamster and less like the most sensitive guinea pig.

DR. FIRESTONE: Has the protein receptor been identified?

DR. NEAL: The receptor which has been tentatively identified is a very labile protein of 150,000 molecular weight, which is in the cytosol portion.

DR. KOLBYE: Does phenobarbital treatment protect animals against TCDD?

DR. NEAL: Phenobarbital or TCDD pretreatment of the rat raises the LD_{50} about equally. There does not appear to be any advantage in administering phenobarbital to a person who has been exposed to TCDD in terms of any competitive binding affinity for the hepatocytes.

CHAPTER 5

MICROBIAL DEGRADATION OF TCDD IN A MODEL ECOSYSTEM

F. Matsumura, John Quensen, and
G. Tsushimoto

Pesticide Research Center
Michigan State University
East Lansing, Michigan

I. INTRODUCTION

Since it was discovered that TCDD was a contaminant in the widely used herbicide 2,4,5-T (2,4,5-trichlorophenoxyacetic acid)(Matthiasch 1978), it has become a matter of great concern of many scientists. Whether this type of chemical persists in the environment for long periods has stimulated many studies. There are several reports predicting the overall persistence of TCDD using model materials and ecosystems. Kearney et al. (1973), for instance, reported that TCDD was persistent in soil under certain conditions with a half-life of about 1 year. Crosby and Wong (1977) showed that TCDD could photochemically degrade at a relatively fast rate. According to the report by Matsumura and Benezet (1973), TCDD was almost immobile in the soil, remaining on the top-most layer. They concluded that TCDD, because of its poor lipid solubility, is not likely to bioaccumulate in a biological system as does DDT. Isensee and Jones (1975), on the other hand, concluded that TCDD accumulates in fish, which is based on the amount

ISBN 0-12-193160-9

of TCDD available in water. In their model ecosystem, degradation of TCDD did not begin for 1 month. Using a model aquatic system, Ward and Matsumura (1978) showed that most of the TCDD deposited on the sediment. Small amounts of TCDD metabolites from the sediment were also released into the water.

The bulk of TCDD in the environment is expected to end up in soil and aquatic sediment as shown in the case of DDT, PCB, and other organic pollutants. Therefore, a question must be raised whether TCDD is degraded by soil microorganisms, and if so whether there is any way to stimulate such degradation activities to lower its residue levels. To answer the question we have decided to first investigate microbial degradation in the environment. Also we have studied the characteristics of such microbial degradation of TCDD by using model ecosystems and defined microbial cultures. The results are reported below.

II. MATERIALS AND METHODS

The fate of TCDD in an aquatic ecosystem was studied under two sets of experimental conditions. The first was an outdoor pond experiment, while the second was an indoor model ecosystem.

The outdoor pond was an artificial circular pool with a radius of 6.85 m, lined with 30 cm-thick concrete wall. The depth of the pond water was initially 1.1 m and increased to 1.4 m one year later. Another nearby pond identical in size and structure (approximately 20 m west) served as a control. The sediment consisted of a top layer of 5 cm of clay organic loam, and a bottom layer of 50 cm of sand. The composition of the top layer was sand 79.3%, silt 0%, clay 20.7% with 4.1% organic matter (0.75 ppm nitrate) at a pH of 7.6. The predominant pondweed species were *Elodea nuttali* and *Ceratophyllon demersum*. The biomass of pondweeds was estimated to be about 813 g/m^3 (fresh weight)(October 18, 1978). The fish used for the experiment was the fathead minnow. For accumulation studies 50 fish were placed in two separate cages and dipped in the upper water layer about 40 cm below the surface. On October 18, 1978, 3.4 mCi ^{14}C-TCDD (8.7 mg) in 6.8 ml of anisole and 1 ml of Triton X-100 was added to 20 liters of pondwater. The solution was thoroughly mixed by shaking and spread over the pond surface homogeneously. The specific activity of TCDD was 126 mCi/mmol (uniformly ring labeled). The initial concentration of TCDD in water was 53.7 ppt. On each sampling day, upper water (0.2-0.3 m below the surface), lower water (1.0-1.1 m below the surface), sediment (from the surface to about 5 cm depth), pondweeds, and fish were collected. For routine assay, the radioactivity in the 0.5 ml water was measured directly by liquid scintillation spectroscopy. To estimate TCDD concentration below the 10 ppt level, 1-12 liters of water sample were extracted with 500 ml of chloroform, dried over sodium sulfate, solvent evaporated, the residues picked up in liquid scintillation counting solution, and the radioactivity assayed. TCDD and its metabolites were extracted from 10 gm of sediment by the series of solvents described by Ward and Matsumura (1978). A preliminary test established that the recovery of the added radioactivity was complete using this method. The recovery values were 95.1% in one experiment and 105% in another. For fish and pondweeds experiments, two fish were collected and weighed, and 2 g of pondweeds were collected. TCDD and its metabolites in fish and pondweeds were extracted by homogenization with 20 to 25 ml of chloroform. In all cases the data were calculated on a fresh weight basis (e.g., ppt) based upon radioactivities and expressed as the amount equivalent to the original concentration of TCDD.

For the metabolic study of TCDD, 5 liters of upper water, 50 gm sediment, and 40 g of pondweeds were collected after 1 year. TCDD and its metabolites in the upper water were extracted with 500 ml of chloroform. In the cases of the sediment and pondweed samples, 150 ml of acetone and 400 ml of chloroform were used, respectively. The solvent extracts were evaporated to dryness; and the residues were picked up in 2 ml aliquots of chloroform (in the case of water and weed), or acetone (in the case of sediment) for thin layer chromatographic (TLC) analysis.

Polar metabolites were separated from TCDD on the TLC plate (LK5F-TLC, Whatman) by using carbon tetrachloride as a mobile phase (Matsumura and Benezet, 1973). TCDD on the plate was scraped from the area of R_f 0.36-1.0 and polar metabolites were similarly collected from the area of R_f 0-0.1. Radioactivity in the silica gel was measured by combustion and liquid scintillation counting.

The model ecosystem experiments were carried out in a glass bottle (radius: 16 cm, height: 23.5 cm), each containing 1100 gm of sediment, 2.3 liters of water, 10 gm of pondweed, and one fish (0.5-1.0 gm fish). These materials had been taken from the same outdoor pond as the first experiment before the experiment started. The model ecosystems were kept on the laboratory bench with a glass cover and a 100 w florescent light source about 40 cm above the top of the water surface. On the first day of the experiment, ^{14}C-TCDD of specific activity 126 mCi/mmol (uniformly ring labeled) was added to make final concentration of 61.36 ppt. On each sampling day 10 gm of sediment, 10 ml of water, 1 gm of pondweed (fresh weight), and one fish were collected. Radioactivities were measured by the same methods as used for the outdoor pond experiment. The model ecosystem experiment was continued up to 45 days.

The details of the design of the aquatic microbial ecosystem were described by Ward and Matsumura (1978). The system consisted of the sediment and the lake water kept in a loosely capped 20-ml culture tube in darkness at room temperature.

The terrestrial microbial ecosystem consisted of, respectively, 2 cm of glass wool, 25 gm of washed, ignited sea sand (Mallinckrodt, Inc.) and 20 gm of sieved soil, added to a 50-ml Pyrex beaker. The soil was collected from one of the areas on or near the MSU campus and represented farm, garden, and oak woods types. Five milligrams of finely ground naphthalene was gently stirred into the soil in half of the beakers. This gave a concentration of 250 ppm of naphthalene in the soil. The soil "plugs" were then covered with aluminum foil to exclude light, but not air, and placed in a growth chamber at 30°C with 95% RH. Soil moisture was visually checked each week, and distilled water was added as needed to maintain the soil in a moist but unsaturated state. After 2 weeks, 1 ml of a 0.05 mCi/ml DMSO solution of TCDD was added to each beaker of soil. Extractions were made 2 and 4 months after the addition of TCDD.

Soil, sand, and glass wool were extracted as follows: once with 100 ml chloroform:methanol (2:1), then with 50 ml chloroform:methanol, and finally, with 50 ml chloroform. All three extracts were combined, evaporated, and the residue redissolved in 0.4 ml ethyl acetate.

Separation of TCDD and metabolites was by TLC (LK5D-TLC, Whatman) with carbon tetrachloride as the mobile phase. Radioactivity on the plates was visualized by scanning and/or autoradiography, and the appropriate areas scraped and counted by liquid scintillation.

III. RESULTS

A. *COMPARISON OF THE FATE OF TCDD IN THE OUTDOOR POND AND IN THE MODEL ECOSYSTEM*

The fate of TCDD in the aquatic system was studied using an outdoor pond. The result was compared with that of model ecosystem studies. As shown in Figs. 1 and 2, TCDD was distributed in the sediment, water, pondweed, and fish in the outdoor ecosystem. The initial buildup of TCDD in the sediment reached 2700 ppt. The radioactivity in the sediment gradually decreased from 2700 to 500 ppt within 50 days (Fig. 1). The bioaccumulation of TCDD in the fish and pondweed proceeded rapidly, as shown in Fig. 2. Accumulation of radioactivity by

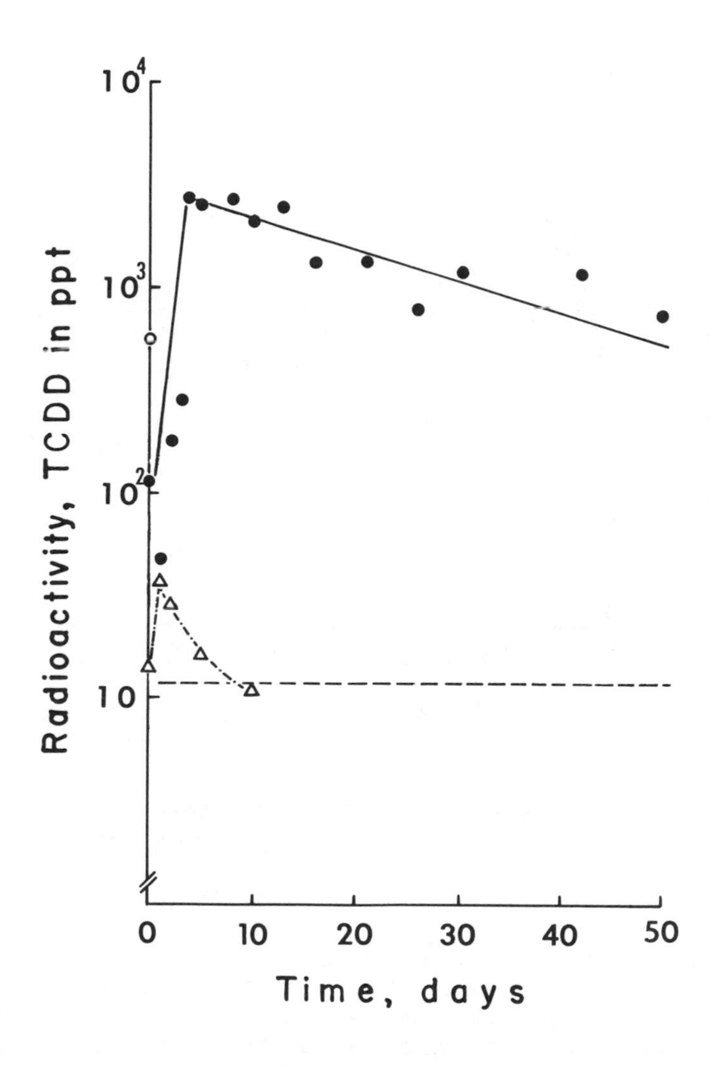

Fig. 1. Short-term distribution of TCDD-derived radioactivity among the sediment (closed circles), upper layer (open circles), and lower layer of water (open triangles) in the outdoor pond experiment. The data are expressed in the equivalent of the originally added TCDD as assessed by the radioactivity. Dotted line shows a detection limit. (Tsushimoto et al., 1982. Reprinted with permission of SETAC, Publishers.)

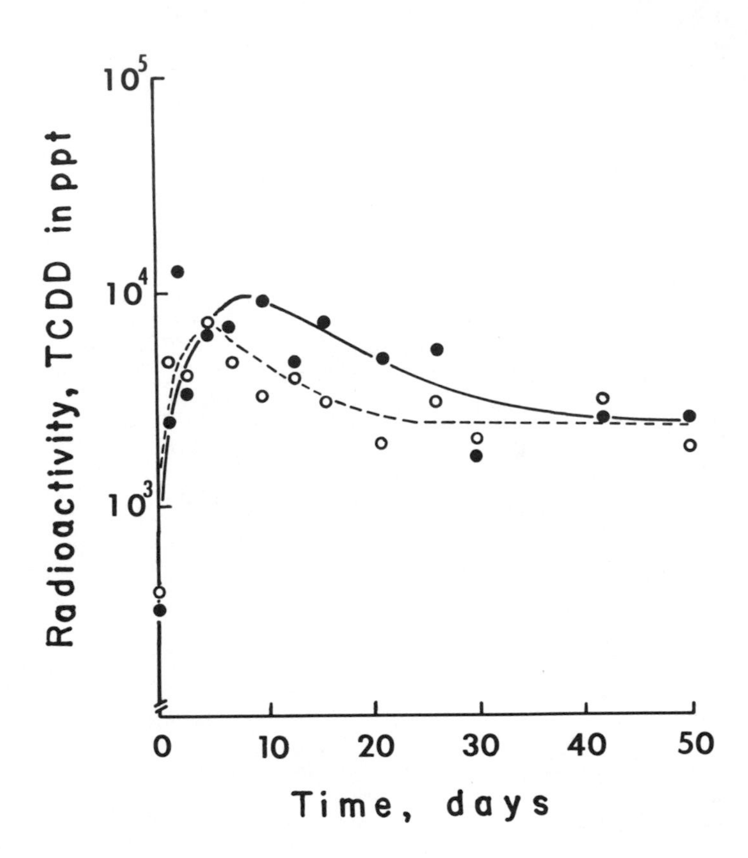

Fig. 2. Short-term accumulation of TCDD-derived radioactivity by fish (closed circles) and pondweed (open circles) in the outdoor pond experiment. (Tsushimoto et al., 1982. Reprinted with permission of SETAC, Publishers.)

the fathead minnow increased for about 10 days, and decreased thereafter to reach a constant level after 40 days (2500 ppt). The maximum accumulation level in the minnow was 8500 ppt after 10 days. The bioaccumulation of the radioactivity by the pondweeds reached a maximum level (7000 ppt) after 5 days. At the time of the initial equilibrium (1 month), the level of accumulation in the pondweed was on the order of 2500 ppt based upon an assumption that the entire radioactivity was due to TCDD.

Similar results were observed with TCDD in the model ecosystem study in the sediment and fish (Figs. 3 and 4). However, the level of bioaccumulation of TCDD by the fish in the bottle was somewhat higher than that found in the outdoor

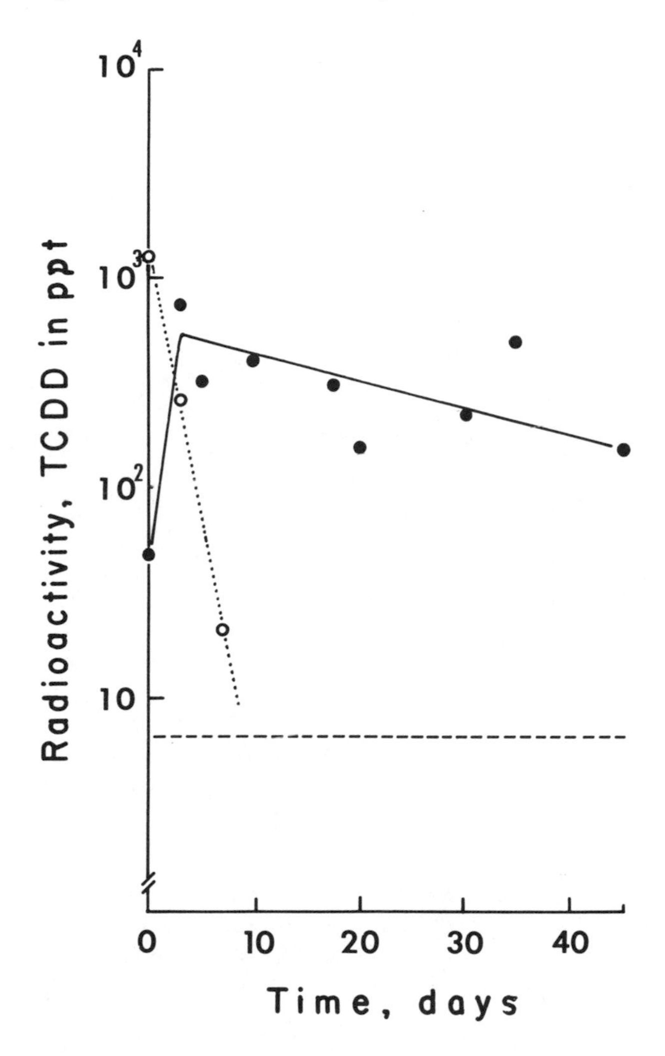

Fig. 3. Distribution of TCDD-derived radioactivity between water (open circles) and sediment (closed circles) in the model ecosystem experiment. Dotted line shows a detection limit. (Tsushimoto et al., 1982. Reprinted with permission of SETAC, Publishers.)

pond experiment. Maximum bioaccumulation of TCDD in fish in our experimental conditions was 5 days.

In the case of pondweed in the model ecosystem experiment, the level of radioactivity was highest at the start of the experiment. Thereafter, it decreased, reaching a constant

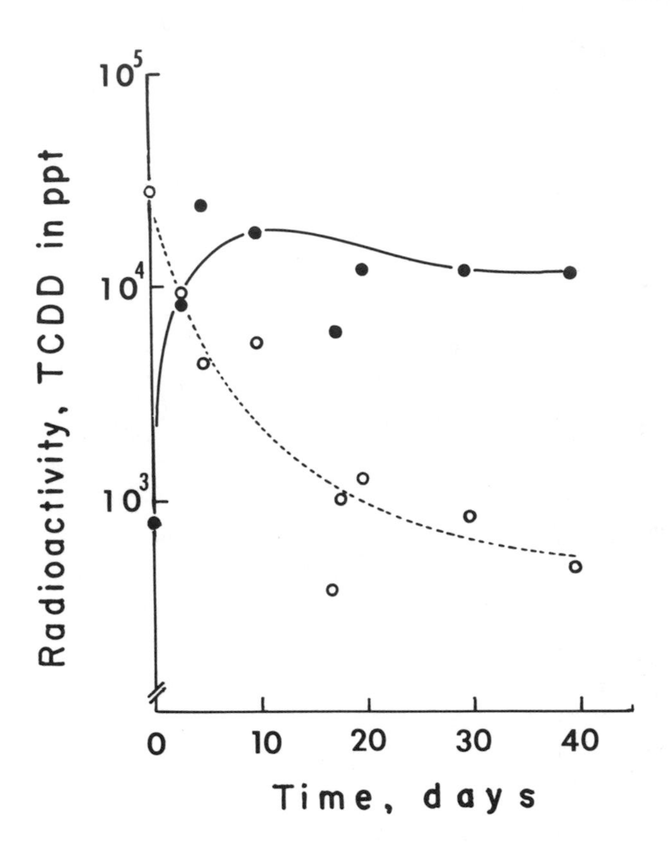

Fig. 4. Bioaccumulation of TCDD-derived radioactivity in fish (closed circles) and pondweed (open circles), in the model ecosystem experiment. (Tsushimoto et al., 1982. Reprinted with permission of SETAC, Publishers.)

level (500 ppt) in about 1 month. This mode of accumulation was different from the data obtained in the outdoor pond experiment. Also the level of TCDD accumulation was much lower in the model ecosystem than that obtained in the outdoor pond.

The results of the 65 days to 1 year monitoring study on accumulation of TCDD in the sediment, pondweed, and fish indicate that the level of accumulation of TCDD into each component reached a constant level under this experimental condition (Figs. 5 and 6). The levels in the biological systems were as follows: sediment 70 dpm/gm (80 ppt if all radioactivity was derived from entire TCDD), pondweed 1740 dpm/gm (2000 ppt), and fish 2170 dpm/gm (2500 ppt), respectively. The level of radioactivity in the sediment reached a constant in about 6 months under the experimental condition.

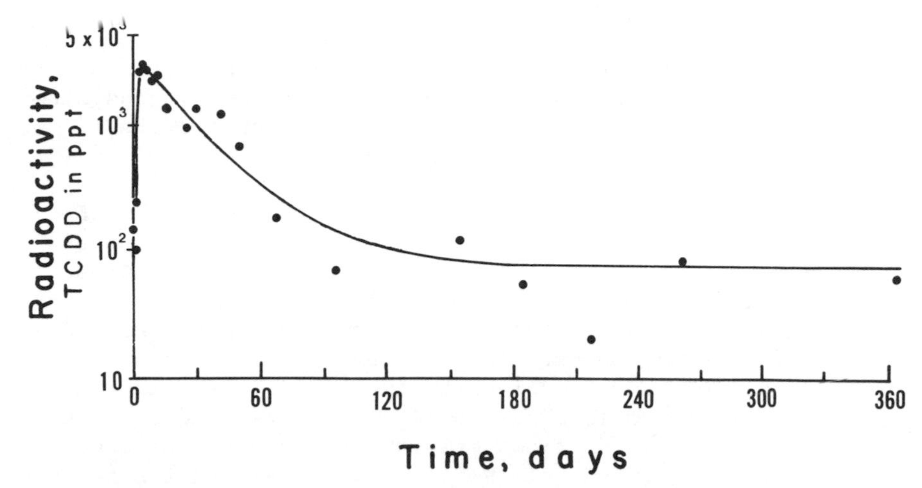

Fig. 5. Long-term distribution of TCDD-derived radioactivity among sediment in the outdoor pond experiment. (Tsushimoto et al., 1982. Reprinted with permission of SETAC, Publishers.)

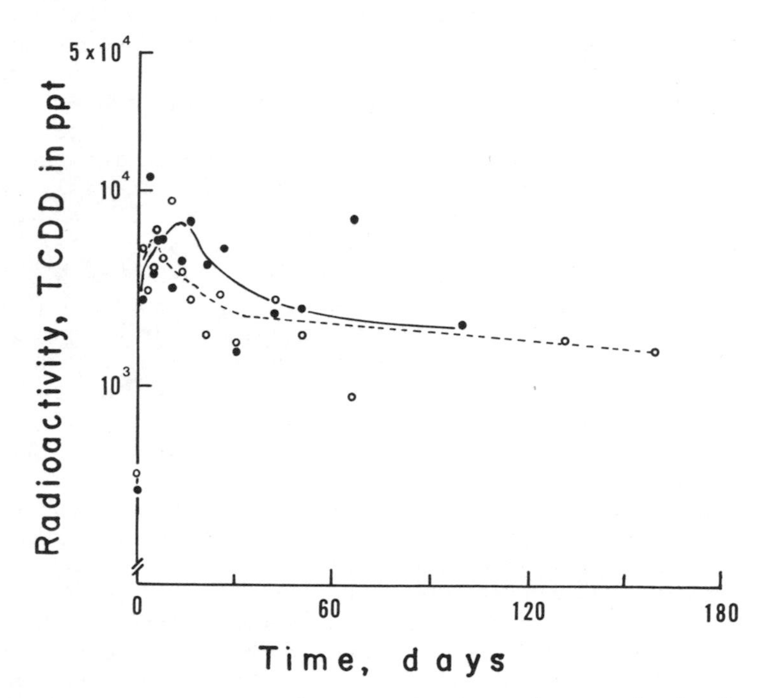

Fig. 6. Long-term bioaccumulation of TCDD-derived radioactivity in fish (closed circles) and pondweed (open circles) in the outdoor pond experiment. (Tsushimoto et al., 1982. Reprinted with permission of SETAC, Publishers.)

The results of a general survey on the distribution pattern of TCDD is summarized in Table I. It can be seen that at 40 to 50 days, the data obtained by using the model ecosystem were roughly comparable to the ones from the outdoor test. One exception was the data for the pondweed.

Metabolic changes of ^{14}C-TCDD were then studied. The ratio of TCDD and its metabolites after 1 year were obtained; the results are shown in Table II. At this time, the water contained 60% of the radioactivity as TCDD and the rest as polar metabolites (40%). On the other hand, almost all of the radioactivity in sediment was due to unmetabolized TCDD. This result confirms the observation made by Ward and Matsumura (1978) that TCDD is metabolized in the sediment and the metabolites are released into water. It is interesting to note that the level of metabolites present in pondweed was almost identical to that found in water. It is likely that both TCDD and its metabolites are distributed by similar partition coefficients between water and pondweed.

At the end of 1 year an effort was made to survey the amount of TCDD left in the system. To carry out the task the total weight of each component was estimated. In the case of the sediment, TCDD was found to be exclusively present in the topmost organic soil layer. The average depth of the layer was obtained by sampling several stations. The sediment samples were obtained by using a clear acrylamide tube (2 cm i.d.). The top layer had a distinct dark color. This portion was carefully collected apart from other soil materials. Within the top layer the sample was mixed to obtain the average TCDD concentration. The samples were spread over several layers of paper towels and air dried for 1 day. The weights of the air-dried samples were taken for the calculation of TCDD levels. The moisture content of the air-dried sample was 13.5 ± 4.3% as judged by oven drying overnight at 200°C.

The results shown in Table III clearly show that the bulk of TCDD is tied up in pondweed. The total recovery of unmetabolized TCDD was 49.7% of the initial added quantity. At the end of a 25-month period, the same survey effort was made to assess the amount of remaining TCDD. The most noticeable change in the pond from the 1 year posttreatment observation time was that a massive death of the pondweed resulted in the accumulation of decayed plant matter at the bottom of the pond. This extra layer was clearly distinguishable from the top organic soil layer. Both layers were found to contain radioactivity. They were radioassayed separately, but the results are expressed as a combined figure (Table IV). As a result of sedimentation of the dead plant matter, the bulk of radioactivity was found to have shifted from the pondweed to the sediment component. An intriguing observation is that the

TABLE I. Levels of Total Radioactivity Expressed in TCDD Equivalent[a]

Source	Incubation time (days)	TCDD equivalent (ppt)			
		Water	Sediment	Fish	Pondweed
Outdoor	50	<12	500	2500	2500
Outdoor	365	0.066(0.026)[b]	97[c](3.5)	---	2000(794)
Outdoor	760	<0.0043	321[d](9.6)	---	8.8(3.6)
Indoor	40	<12	200	2000	500

[a] *Tsushimoto et al., 1982. Reprinted with permission of SETAC, Publishers.*

[b] *Metabolites are shown in parentheses.*

[c] *This value was obtained in the top organic layer.*

[d] *The increase is due to death of pondweed and resulting accumulation of decayed plant matter on the surface of the top layer.*

TABLE II. Ratio between TCDD and its Metabolites after 1 Year Pond Experiment[a]

	Upper water (%)	*Sediment* (%)	*Pondweed*[c] (%)
TCDD[b]	60.2%	96.4%	60.3%
Metabolites[b]	39.8%	3.6%	39.7%

[a]*Tsushimoto et al., 1982. Reprinted with permission of SETAC, Publishers.*

[b]*TCDD and its metabolites were isolated with chloroform in the case of water and pondweed, and with acetone in the case of sediment. These were analyzed on the TLC plate.*

[c]*Pondweed was collected after 14 months.*

TABLE III. Estimation of Total TCDD Remaining in Various Components after 1 Year[a]

	Volume and weight	*Radioactivity*	*Unmetabolized TCDD (%)*	*Amount of TCCD (ppt)*[b]	*Total amount of TCDD remaining*
Water	206 m^3	0.057 dpm/ml	60.2	0.04	8.2 μg
Pondweed	2.72 t	2000 dpm/gm[d]	60.3	1390	3.78 mg
Sediment	5.75t[c]	84 dpm/gm	96.9	94	0.54 mg
					4.32 mg[e]

[a] *Tsushimoto et al., 1982. Reprinted with permission of SETAC, Publishers.*

[b] *1 dpm/gm TCDD was equivalent to 1.15 ppt. To obtain the value for water it was necessary to extract 5 liters of water with chloroform.*

[c] *This value was estimated by the weight of the top organic soil layer.*

[d] *This value was estimated by using Fig. 6.*

[e] *This value was 49.7% of initial TCDD added.*

TABLE IV. Estimation of Total TCDD Remaining in Various Components after 25 Months[a]

	Volume and weight	Radioactivity	Unmetabolized TCDD (%)	Amount of TCDD (ppt)[b]	Total amount of TCDD remaining
Water	249 m^3	0.0037 dpm/ml	---	4.3×10^{-3}	1.07 μg[d]
Pondweed	0.1 t	7.7 dpm/gm	59.5	5.2	0.52 μg
Sediment[c]	8.47 t	279 dpm/gm	97.0	311	2.56 mg[e]
					2.56 mg

[a] *Tsushimoto et al., 1982. Reprinted with permission of SETAC, Publishers.*

[b] *1 dpm/gm TCDD was equivalent to 1.15 ppt.*

[c] *This sediment contained decayed plant matter and the top organic soil layer. The former layer formed as a result of a massive death of pondweed which took place sometime between the 12th to 25th month posttreatment.*

[d] *Because of the low level of radioactivity the ratio of metabolites to TCDD could not be determined. The values shown assume that the entire radioactivity was due to TCDD and, hence, are overestimated.*

[e] *This value was 29.4% of initial TCDD added.*

newly formed sediment did not contain a high proportion of polar metabolites. A calculation would show that if the entire unmetabolized TCDD in the pondweed from the previous year sedimented and added to the already existing TCDD in the top organic soil layer it would give 2.8 mg of total TCDD. This value is not too far above the actual figure (2.56 mg) found at the end of the second year. The total recovery of TCDD for the second year was 29.4%. These recovery figures are generally lower than the ones produced by Ward and Matsumura (1978) using a model system. However, at this stage we cannot assure the dissipation of TCDD according to first-order kinetics.

In summary, the 53.7 ppt of TCDD added to the water slowly dissipated from the outdoor pond. The amounts of TCDD found after 1 and 2 years were 49.7 and 29.4%, respectively. The model laboratory system using the identical components from the outdoor pond produced roughly comparable data for short-term distribution behavior. The greatest change was in pondweed. This material produced the largest biomass changes due to rapid growth and subsequent massive death, causing a large shift in TCDD distribution and probably an anaerobic environment. While the source of metabolic activity has not been determined, the bulk of TCDD metabolites was detected in water and algae. Metabolites disappear from the system faster than TCDD itself as attested to the decrease in the overall level of metabolites in the system in the second year. In this case, evaporation is the likely cause.

As for the difference in the TCDD persistence figures between the laboratory-produced data (Ward and Matsumura, 1978) and outdoor ones, no solid evidence singled out the cause, since many factors could interact in the outdoor environment. If one is allowed to speculate the most likely major cause for the difference, one could select the effect of sunlight combined with the influence of algae. Ward and Matsumura (1978) placed the system under ordinary laboratory light in a thick Pyrex glass container. In the outdoor environment sunlight intensity is, of course, much stronger. Combined with the knowledge that TCDD is relatively labile to photochemical attack and algae are known to produce photosensitizers (Matsumura and Esaac, 1979), it would not be too surprising to find that sunlight contributes significantly to dissipation of TCDD in the outdoor environment, making the half-life figure less for the outdoor experiment as compared to the ones derived in the laboratory without photochemical simulation devices.

B. AQUATIC AND TERRESTRIAL MICROBIAL MODEL ECOSYSTEMS

The above study has established that TCDD, although a very stable chemical, is still slowly degraded in the environment.

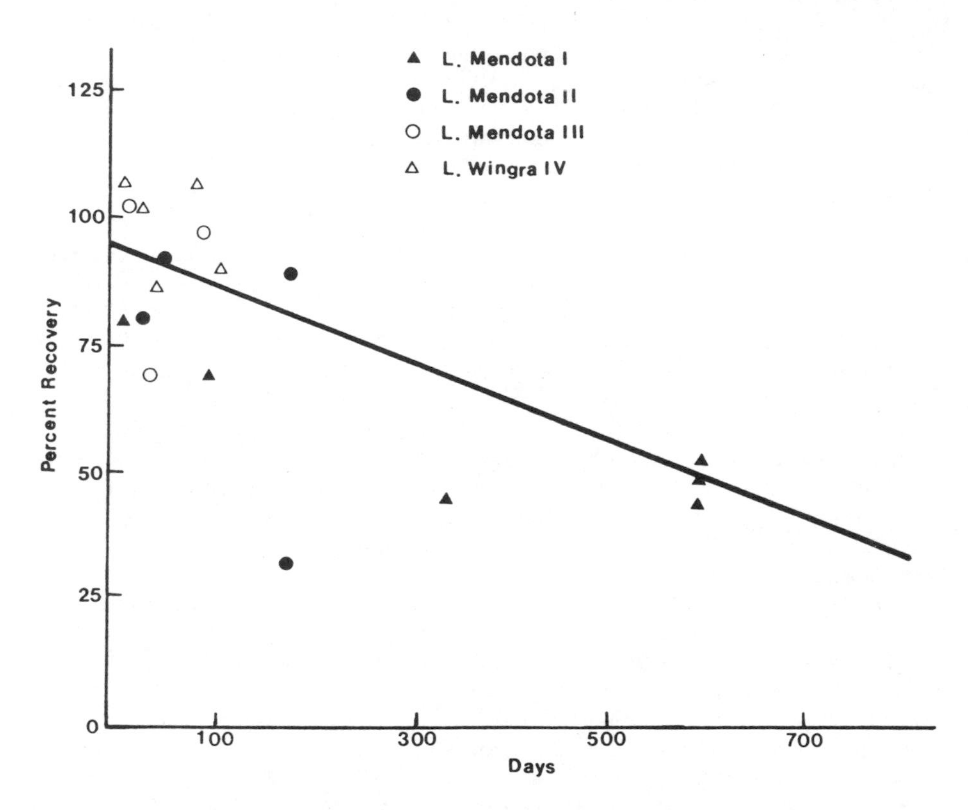

Fig. 7. Percentage recovery of TCDD from lake sediments in relation to time of incubation: the correlation coefficient of the line is 0.731. (Ward and Matsumura, 1978. Reprinted with permission of Springer-Verlag, Publishers.)

In order to demonstrate that microbial activities are important in this regard, two sets of microbial model ecosystems were designed. The first system was an aquatic one. For this system, large quantities of sediment samples were obtained from University Bay, Lake Mendota, and Lake Wingra, Madison, Wisconsin and stored in a cold room at 4°C. For incubation tests, 5 gm (wet weight) of the sediment and 18 ml of lake water were added to each 20-ml culture tube, shaken, and equilibrated. ^{14}C-TCDD was added and the screw-capped tubes kept at room temperature for a specified time period. The results summarized in Fig. 7 show that there is an overall decline in the level of TCDD recovered from the ecosystem. It must be pointed out that the extraction efficiency of this test series was rather good, the range of recovery being 92-105% (Ward and Matsumura, 1978).

To ascertain that the microbial activities contribute to the overall disappearance of TCDD from the system, two types of nutrients, glucose and bactopeptone, were added to the system, and their influence on TCDD metabolism studied (Table V). Despite the variation which is expected for this type of study, the results indicate that the level of total metabolites formed in the aqueous phase of the systems which contained added nutrients are higher than those found in unaltered systems. It must be pointed out that in the above study only the radiocarbons found in the aqueous phase were analyzed (i.e., those in the sediment were not studied). Earlier it was noted that the ratio of metabolites to TCDD was much higher in the water phase than that found in the sediment. Yet, when TCDD was incubated with filtered lake water, in the absence of sediment, metabolic products did not form (Ward and Matsumura, 1978), indicating the source of microbial activities in sediment. This phenomenon has been interpreted to mean that the metabolic products, having higher water solubilities, were released to water from the sediment. This interpretation agrees well with the observation made earlier in the outdoor pond experiment. For comparison the overall metabolic patterns involved in the entire model ecosystem were studied. The results are shown in Table VI. Samples II34 and II32 in this study are identical to the ones listed in Table V. Here the percentage of the metabolic products found was much lower than that found in the aqueous phase because of the inclusion of a higher level of unmetabolized TCDD in the sediment.

Another factor which might be contributing to the overall loss of radiocarbon is evaporation. In this series of experiments some water losses were noted, particularly in those systems that were kept over 1 year. The reason for this loss of water is that the screw caps of the tubes were not completely tightened in this model ecosystem to avoid explosion due to formation of gas during the incubation period. When the loss of water through evaporation was plotted against the percentage recovery of TCDD (Fig. 8), a general correlation between these two parameters was observed.

A model terrestrial, microbial ecosystem consisting of a soil plug (20 gm) placed in a 50-ml beaker over a 2 cm layer of glass wool and a 25 gm equivalent layer of washed ignited sand was set up. The system was kept under moist conditions as was outlined in an earlier section. Three different types of soil samples were used in the presence and absence of 250 ppm of naphthalene, which was added to promote microbial activities to metabolize aromatic compounds. After 2 and 4 months of incubation the soil samples were extracted. The remaining unmetabolized ^{14}C-TCDD was assessed by using thin-layer chromatography (Table VII). The result indicates that

TABLE V. Metabolite Formation in the Aqueous Phase of Samples Incubated with Sediment and Water[a]

Sample	*Days incubated*	*Radioactivity recovered*[b] *(%)*	*TLC distribution of radioactivity*	
			TCDD (%)	*Total metabolites (%)*
Glucose				
II40	0	5.49	90.00	10.00
II26[c]	19	0.54	51.70	48.29
II32[c]	39	2.40	39.36	60.64
II26	163	1.34	1.73	98.27
II32	167	1.38	47.40	52.60
Bactopeptone				
II41	0	1.20	94.14	6.14
II27[c]	19	3.22	22.37	77.63
II33[c]	39	0.54	13.85	86.15
II27	163	4.84	63.87	36.13
II33	167	1.41	11.08	88.92
Glucose + bactopeptone				
II28[c]	19	9.38	30.18	69.82
II34[c]	39	1.20	14.28	85.71
II34	167	1.78	43.09	56.90
No nutrients				
II43	0	3.31	89.63	10.36
II29[c]	19	0.94	54.63	45.37
II35[c]	39	4.07	66.67	33.33
II35	167	0.32	54.25	45.74

[a] *Ward and Matsumura, 1978. Reprinted with permission of Springer-Verlag, Publishers.*

[b] *Percentage administered radioactivity that was recovered in the aqueous phase.*

[c] *Water removed via freeze-drying. Water in the other samples was removed by evaporation.*

TABLE VI. Metabolite Formation in the Sediment [a]

Sediment sample	Days incubated	TLC distribution of radioactivity[b]			Nutrients[c]
		% TCDD	Total metabolite (5)	Total recovery (%)	
III60	0	92.52	7.48	101.78	--
IV66	0	93.67	6.33	102.14	--
II28	19	96.19	3.81	53.28	G,B
IV64	22	98.35	1.65	103.28	--
II34	39	91.71	8.29	91.36	G,B
IV64	67	97.47	2.53	107.96	--
IV64	98	89.04	10.96	90.03	--
II32	167	97.48	2.52	89.03	G
I1	333	76.78	23.22	43.09	--
I22	586	98.96	1.04	50.41	--
I11	588	98.74	1.26	44.47	--
I17	588	95.59	4.41	52.15	B

[a] *Ward and Matsumura, 1978. Reprinted with permission of Springer-Verlag, Publishers.*

[b] *Expressed as percentage of radioactivity found in TCDD position and lower R_f positions on TLC plate, respectively.*

[c] *G, glucose; B, bactopeptone.*

TABLE VII. Results after 2 and 4 Months Incubation of Soil Samples, with and without Addition of 250 ppm Naphthalene

Soil	*Months*	*Aqueous phase*[a]	*Solvent phase*	
			Metabolite	*TCDD*
Woods	2	--[b]	0.6	54.8
	4	0.3	1.1	72.0
Woods-naphthalene	2	--	0.9	68.9
	4	2.1	0.9	56.6
Farm	2	--	1.2	72.2
	4	28.3	0.1	0.4
Farm-naphthalene	2	--	0.9	78.9
	4	0.9	3.0	64.8
Garden	2	--	0.8	68.1
	4	5.8	0.7	61.6
Garden-naphthalene	2	--	0.9	91.6
	4	1.8	1.1	75.8

[a] *Moisture originally in soil, left after extraction and evaporation of solvent.*

[b] *The samples were not analyzed.*

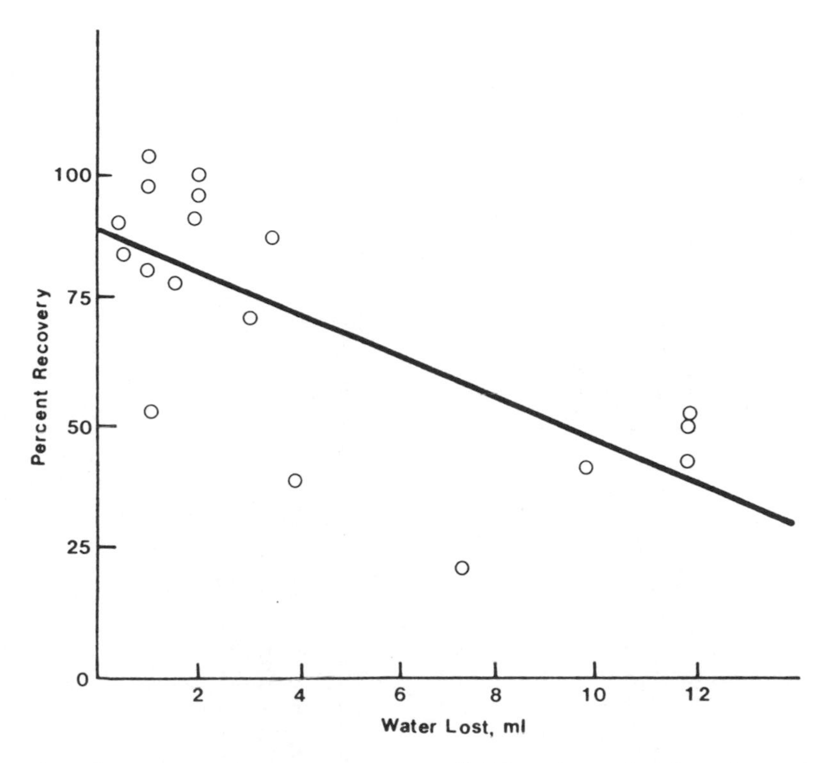

Fig. 8. Percentage recovery of TCDD in relation to water loss from the incubation container: the correlation coefficient of the line is 0.716. (Ward and Matsumura, 1978. Reprinted with permission of Springer-Verlag, Publishers.)

after 4 months approximately 66% of the originally added TCDD remained in soil. Naphthalene did not have a significant effect on the rate of soil degradation of TCDD. As a result of thin-layer chromatographic analysis of these samples some polar metabolic products were recognized. These two sets of experiments have established that TCDD does indeed degrade in aquatic sediment and in soil. However, to prove that microorganisms are capable of degrading TCDD, the following tests using microbial cultures were conducted. In the experiment summarized in Table VIII, 1 gm of sample from each type of soil was added to the nutrient medium and incubated for 7 days. The samples were extracted and analyzed as before. As judged by the level of solvent-soluble metabolites formed, the addition of 40% of No. 2 medium had a stimulatory effect on TCDD metabolism. The microbial isolates in this medium also gave

TABLE VIII. Effects of Nutrient Media Added to Soil Samples on the Metabolic Activities of TCDD[a]

Treatments (number of experiment)	*Naphthalene (ppm)*	*Aqueous phase*	*Solvent phase*	
			Metabolites	*TCDD*
Soil[b] plus No. 2[c]medium (10)	100	4.0 ± 1.9	1.5 ± 1.1	58.1 ± 5.9
Soil plus basal salt (4)	1000	4.7 ± 4.1	0.2 ± 0.1	39.3 ± 9.9
Microbial isolates in No. 2 medium (10)	100	9.7 ± 5.4	0.9 ± 0.3	62.1 ± 5.8
Microbial isolates in basal salt (8)	1000	3.5 ± 4.2	0.3 ± 0.1	74.5 ±14.9

[a]*As controls microbial isolates were obtained by using respective media transferred to fresh media after 1 day and incubated with TCDD for 7 days.*

[b]*Soil samples used were four samples from mixed woods land, two dried river bottom sample, and four farm soil samples. Incubation period was 14 days.*

[c]*40% No. 2 medium (yeast-mannitol) plus naphthalene as indicated.*

a higher metabolite production level than in the basal salt medium.

C. *METABOLISM OF TCDD BY DEFINED MICROBIAL ISOLATES*

A preliminary screening test established that among several isolates tested, *Nocardiopsis* sp. (Beeman and Matsumura, 1981) and *Bacillus megaterium* (ATCC No. 13368) showed some TCDD-degrading activities. In the first experiment an effort was made to assess the effect of naphthalene, which was added to the medium as an alternate carbon source, and mannitol. The results shown in Table IX indicate that there is a slight increase in the level of radiocarbon in the aqueous phase and a slight decrease in the level of TCDD remaining unmetabolized. The differences, however, were statistically insignificant. In the next experiment the effect of decreasing the level of nutrient and replacement with the basal medium (i.e., salts without any carbon source) containing 10 ppm of naphthalene was assessed. The results (Table X) indicate that the overall metabolic activity as judged by the level of solvent soluble metabolites found through thin-layer chromatographic analyses was higher, when there was some amount of mannitol-yeast medium present than in the case where naphthalene was the sole carbon source. At this stage, the effect of carrier solvent, which was used to dissolve TCDD and used as a vehicle for administration, was reexamined. The results (Table XI) indicate that both ethyl acetate and DMSO are better carriers than ethanol or corn oil as judged by the solvent-soluble metabolites and the amount of radiocarbons in the aqueous phases found. By using the same criteria ethyl acetate was judged as the better of the two. An experiment used to confirm this carrier effect was conducted with varying levels of soybean extract in the medium (Table XII). The result clearly shows that ethyl acetate is a superior carrier for TCDD, inducing a higher level of TCDD degradation than DMSO. The level of soybean extract did affect the metabolism, particularly in the case of the ethyl acetate-treated sample, where the level of solvent-soluble metabolites increased when the content of soybean extract was reduced from 100 to below 50%. An autoradiogram of the thin layer chromatogram of the solvent phase from one replication of this experiment is shown in Fig. 9. It shows that metabolism of TCDD is extensive in the ethyl acetate-containing samples and that several metabolic products are separated by this type of chromatography.

TABLE IX. Effect of Varying Naphthalene Concentration on Metabolism of TCDD by *Nocardiopsis*[a]

Treatment	*Percentage of ^{14}C activity recovered*		
	Aqueous phase	*Solvent phase*	
		Metabolites	*TCDD*
Naphthalene			
0 ppm	1.9 ± 0.6	5.2 ± 2.3	88.8 ± 9.1
10 ppm	1.7 ± 1.0	3.5 ± 0.8	82.9 ± 5.7
50 ppm	2.1 ± 0.6	3.5 ± 1.7	84.8 ± 5.7
100 ppm	2.4 ± 1.2	3.3 ± 2.0	80.1 ± 6.4
500 ppm	2.4 ± 0.1	6.3 ± 2.7	79.0 ± 4.1
Standard	0	0.6 ± 0.3	99.3 ± 0.3

[a]*Twenty percent No. 2 Mannitol-yeast medium was used in all treatments. Results are expressed as means and standard deviations of three replicates.*

TABLE X. Effects of Culture Medium on Metabolism of TCDD by *Nocardiopsis*[a]

Treatment	Percentage of ^{14}C activity recovered		
	Aqueous phase	Solvent phase	
		Metabolites	*TCDD*
No. 2 medium			
0%	1.5 ± 0.6	4.2 ± 2.0	71.6 ± 6.3
10%	2.0 ± 0.5	7.1 ± 4.1	76.1 ± 6.5
20%	3.0 ± 0.4	8.2 ± 2.7	72.4 ± 5.4
30%	2.8 ± 0.5	7.5 ± 4.1	78.2 ± 2.8
40%	3.1 ± 2.5	7.1 ± 1.4	73.9 ± 9.6
Standard	0	2.6 ± 2.2	97.4 ± 2.2

[a]*Mannitol-yeast medium (No. 2) was diluted to various concentrations with basal medium. Results are expressed as the means and standard deviations of three replicates. Naphthalene 10 ppm was present in all cultures and DMSO was used as the solvent.*

TABLE XI. Effect of Solvents on Metabolism of TCDD by *Bacillus megaterium* in Yeast-Soybean Medium[a]

	Percentage of ^{14}C Activity Recovered			
			Solvent phase	
Solvent	*Aqueous phase A*[b]	*Aqueous phase B*[c]	*Metabolites*	*TCDD*
Ethyl acetate				
1	24.0	3.3	6.4	32.3
2	0.7	5.2	11.3	71.2
DMSO				
1	1.4	2.9	9.5	72.6
2	0.7	0.9	8.2	91.1
Ethanol				
1	0.3	4.2	1.7	94.7
2	2.5	7.5	1.4	85.3
Corn oil				
1	1.7	0.3	--[d]	92.8[d]
2	0.3	0	--	91.2

[a] *The amount of solvent used was 1 ml.*

[b] *Water left after extraction and evaporation of solvents.*

[c] *Aqueous phase after partitioning against the solvent.*

[d] *Thin-layer chromatographic analysis was not possible because of the interference with corn oil.*

TABLE XII. Effect of Culture Medium on Metabolism of TCDD by *Bacillus megaterium*[a]

		Percentage of ^{14}C Activity Recovered	
		Solvent phase	
Treatment	*Aqueous phase*	*Metabolites*	*TCDD*
Ethyl acetate			
100% soybean	10.5 ± 9.4	12.4 ± 7.9	58.7 ± 28.2
50% soybean	16.2 ± 6.7	41.6 ± 10.9	29.6 ± 20.8
25% soybean	18.0 ± 6.8	40.3 ± 13.7	28.2 ± 10.6
12% soybean	16.9 ± 0.6	35.9 ± 2.7	33.3 ± 7.1
(All ethyl acetate samples)	15.3 ± 6.8	32.2 ± 15.5	37.8 ± 22.1
DMSO			
100% soybean	3.3 ± 1.6	13.3 ± 6.4	69.0 ± 4.8
50% soybean	2.1 ± 2.2	24.4 ± 22.7	63.1 ± 23.7
25% soybean	1.8 ± 1.7	8.4 ± 13.3	76.0 ± 7.0
12% soybean	1.0 ± 0.8	0.7 ± 0.5	85.4 ± 7.6
(All DMSO samples)	2.1 ± 1.7	12.7 ± 14.9	72.3 ± 14.1

[a] *Various percentages of the standard amount of soybean extract were used in the yeast-soybean extract media. Results are expressed as the means and standard deviations of three replicates, except two replicates for 12% soybean.*

[b] *Water left after extraction and evaporation of solvents.*

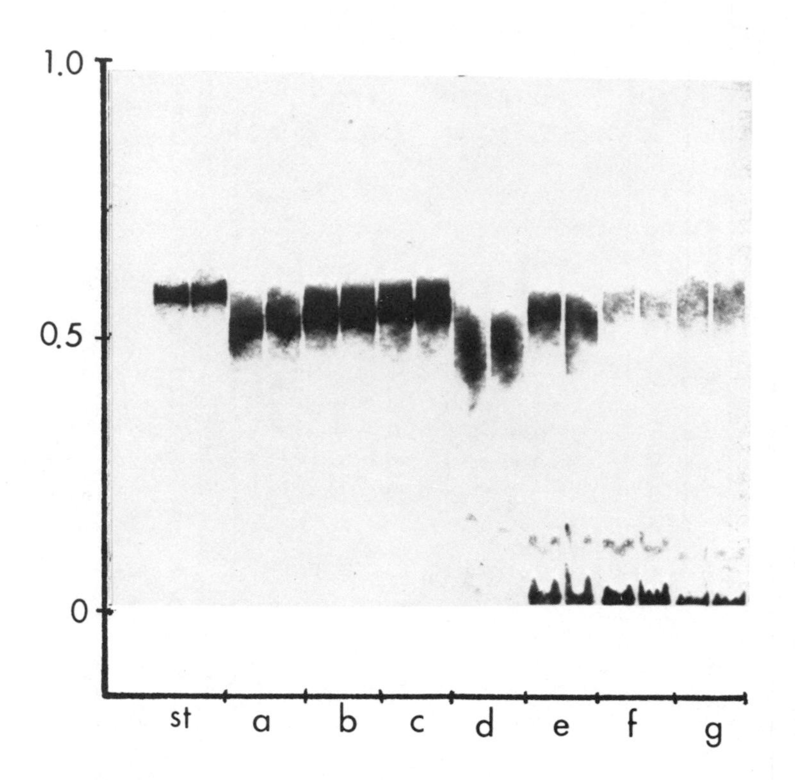

Fig. 9. Autoradiogram of TLC plate of solvent phase for one replication of experiment summarized in Table XII. Bands near the origin (d-g) represent polar metabolites. (st) TCDD standard; (a) DMSO, 50% soybean; (b) DMSO, 25% soybean; (c) DMSO, 12% soybean; (d) ethyl acetate, 100% soybean; (e) ethyl acetate, 50% soybean; (f) ethyl acetate, 25% soybean; (g) ethyl acetate, 12% soybean.

IV. DISCUSSION

There is little doubt that TCDD is relatively stable metabolically and, as a result, stays in the environment for a relatively long period of time. This is in stark contrast to its susceptibility to photochemical reactions, making TCDD a unique environmental contaminant. However, even this microbially recalcitrant pollutant disappears slowly from soil and aquatic environments as shown by our work here and by others (e.g., Kaerney et al., 1973). Three factors considered to be important in this regard are evaporation, photochemical degradation, and microbial degradation. Evaporation is the least studied phenomenon for TCDD, and yet it should be considered as an important process since it makes TCDD available to photochemical degradative forces in the atmosphere. Our preliminary data show that there is a water-mediated evaporation process for TCDD in the aquatic environment. Much more work would be needed, however, to assess the significance of this process, particularly from soil surfaces.

As for the role of microbial activities in degrading TCDD, we have shown both in aquatic sediment and terrestrial soil samples that TCDD is indeed degraded by the action of microbes. The addition of general nutrients, such as, glucose, bactopectone or yeast-mannitol, increases the overall degradation activities. The amounts of metabolic products produced were higher when TCDD was added directly to microbial cultures than in the cases where it was added to soil or sediment. Such a result is expected as the total number of microorganisms in culture are much higher. Also important is the consideration that in soil and sediment some portion of TCDD present is bound to soil constituents and, therefore, is not immediately available to microorganisms.

As to the nature of TCDD metabolism by soil microorganisms, the addition of naphthalene either as an extra or as a sole carbon source in a basal salt medium gave a variable result. Even at the level of pure isolates in culture, naphthalene was stimulatory in one species (*Nocardiopsis* sp.) or variable in another (*B. megaterium*). One possible interpretation of this event is that some strains of microorganisms which have been acclimated to grow on naphthalene as a carbon source also metabolize TCDD, but others could recognize its difference from naphthalene and as a consequence did not degrade TCDD. In this regard it is interesting to note that general nutrients such as glucose, mannitol, and bactopeptone have shown stimulatory properties for microbial degradation of TCDD in all tests conducted so far. At this stage, therefore, one could tentatively conclude that processes involved in such cases are not likely to be very specialized ones. Much more work would be needed to confirm such a possibility.

The most conspicuous stimulatory effect on TCDD metabolism was observed when ethyl acetate was used as a carrier for TCDD. The nature of the carrier appears to be very important (see Table XI) as the use of a wrong carrier, such as corn oil, completely abolished TCDD-metabolizing activity by microorganisms. The most likely explanation of this phenomenon is that TCDD is not readily picked up by microorganisms or hardly penetrates through the microbial cell membrane. TCDD is a unique halogenated aromatic pollutant that has a relatively low lipid solubility (Matsumura and Benezet, 1973). The fact that its bioaccumulation potential is much lower than that of

DDT supports the above hypothesis. Such a difficulty in microbial uptake could be the limiting factor in TCDD metabolism in the environment. This requires much more attention in the future.

V. SUMMARY

Microbial degradation of TCDD was studied by using pure culture isolates of microorganisms, terrestrial and aquatic model systems and an outdoor pond. In each case metabolic activities were recognized by the appearance of metabolic products from ^{14}C-TCDD. In the outdoor pond the apparent half-life of TCDD was in the order of 1 year, recoveries of TCDD after 12 and 25 months being 49.7 and 29.4%, respectively. In model systems metabolic activities on TCDD were stimulated by the addition of general nutrients such as glucose, bactopeptone, and yeast-mannitol. Two microbial isolates, *Bacillus megaterium* and *Nocardiopsis* sp., were found to degrade TCDD. The most important factor found to promote their metabolic activities was the nature of carrier solvent for TCDD. In this regard ethyl acetate gave the best results under the experimental conditions.

ACKNOWLEDGMENT

Supported by the Michigan Agricultural Experiment Station (Journal Article No. 10803), under the Environmental Toxicology Program, and by research grants from Dow Chemical Company, Midland, Michigan and from Syntex, Inc., Palo Alto, California.

REFERENCES

Beeman, R. W. and Matsumura, F. (1981). Metabolism of *cis*- and *trans*-chlordane by a soil microorganism. *J. Agr. Food Chem. 29,* 84-89.

Crosby, D. G. and Wong, A. E. (1977). Environmental degradation of 2,3,7,8-tetrachlorodibenzo-*p*-dioxin (TCDD). *Science 195,* 1337.

Isensee, A. R. and Jones, G. H. (1975). Adsorption and translocation of 2,3,7,8-tetrachloro-*p*-dioxin in an aquatic ecosystem. *Environ. Sci. Technol. 9,* 668.

Kearney, P. C., Woolson, E. A., Isensee, A. R., and Helling, C. S. (1973). Tetrachlorodibenzodioxin in the environment: sources, fate and decontamination. *Environ. Health Perspect. 5,* 273.

Matsumura, F. and Benezet, H. J. (1973). Studies on the bio-accumulation and microbial degradation of 2,3,7,8-tetra-chloro-*p*-dioxin. *Environ. Health Perspect. 5,* 253.

Matsumura, F. and Esaac, E. G. (1979). Degradation of pesti-cides by algae and aquatic microorganisms. *Am. Chem. Soc. Symp. Ser. 99,* 371.

Matthiasch, G. (1978). Survey about toxicological data of 2,3,7,8-tetrachlorodibenzo-*p*-dioxin (TCDD). *In* "Dioxin: Toxicological and Chemical Aspects," (F. Cattabeni, A. Cavellaso, and G. Galli, eds.). SP Medical and Scientific Books, New York and London.

Tsushimoto, Gen, Matsumura, F., and Sago, Ryuichi (1982). *Environ. Toxicol. Chem. 1,* 61-68.

Ward, C. T. and Matsumura, F. (1978). Fate of 2,3,7,8-tetra-chlorodibenzo-*p*-dioxin (TCDD) in a model aquatic environ-ment. *Arch. Environ. Contam. Toxicol. 7,* 349-357.

DISCUSSION

DR. SHEPARD: The ecosystem laboratory set up by Dr. Matsumura is very interesting, as was the demonstration.

DR. AYRES: Because of the low solubility of TCDD in water, would pH influence the rate of reaction of these degradations by the microorganisms?

DR. MATSUMURA: We are limited to the pH range that is possible for the growth of the organisms; this is controlled.

DR. KORTE: Various metabolites were measured. Were carbon-labeled compounds used to measure the ^{14}C-released CO_2? This would be of great ecological significance.

DR. MATSUMURA: No, we have not found any CO_2, although this would be of great interest.

DR. KORTE: We have tried to reproduce your experiment to support your results but we could not. There only seems to be one way to get rid of organic chemicals and this seems to be under atmospheric conditions -- the high energy of the sun and air may be the best and the only way to get rid of these chemicals.

DR. AYRES: The degradation of 2,7-dichlorodibenzo-*p*-dioxin in carbon tetrachloride by the relatively rare oxidative agent ruthenium tetroxide was the basis of our laboratory work.

ISBN 0-12-193160-9

CHAPTER 6

OXIDATIVE CONTROL OF CHEMICAL POLLUTANTS BY RUTHENIUM TETROXIDE

D. C. Ayres

Department of Chemistry
Westfield College
London, England

The Tetroxide is a yellow solid of m.p. 25°, which is freely soluble in organic solvents such as carbon tetrachloride, chloroform, and nitromethane. Its solubility in water at ambient temperature is about 2%, and compares with that of potassium permanganate. It was first prepared by Claus in 1860 by oxidation of the metal with alkaline potassium nitrate. It may be obtained more conveniently in acidic solution by the action of reagents such as ceric ion, bromates, and periodates on the lower valency states of ruthenium; in alkaline solution chlorine or hypochlorite are often employed. The tetroxide may be obtained free from other reagents by steam distillation; solutions in organic solvents are obtained by its generation in an aqueous medium and extraction in a two-phase system, for example,

$$2RuCl_3 + 5ClO^- + 3H_2O \rightarrow 2RuO_4 + 11Cl^-$$

ISBN 0-12-193160-9

Other convenient sources are:

$RuO_2 \cdot H_2O + 2IO_4^- \rightarrow RuO_4 + 2IO_3^-$ (in aqueous acetone)

$RuO_2 \cdot H_2O + 2ClO^- \rightarrow RuO_4 + 2Cl^-$ (in aqueous/CCl_4)

Once an oxidation step is complete the reoxidizable lower oxide is formed and may separate as a black solid:

$$RuO_4 \rightarrow RuO_2 + 2[O]$$

This primary oxidant can, therefore, be used in catalytic amounts while the cheap secondary oxidant is actually consumed.

I. SOME TYPICAL OXIDATIONS (Lee and Van den Engh 1973)

$$R \cdot CH{=}CHR' + 2\,RuO_4 \longrightarrow R \cdot CO_2H + R' \cdot CO_2H + 2\,RuO_2$$

$$\text{(tetrahydrofuran)} + 2[O] \longrightarrow \text{(}\gamma\text{-butyrolactone)} + H_2O$$

In an acidic medium the solvolyzed lactone will undergo further oxidation of the alcohol function

$$\text{(phenyl-R)} \xrightarrow{n[O]} R \cdot CO_2H \quad \text{(no identifiable ring fragments)}$$

Other substrates which resist attack by commercial oxidizing agents are rapidly degraded, for example, chlorobenzene and trichlorethylene. It should be noted that the power of this oxidizing agent limits the choice of solvents to those of electronegative character; even methylene chloride is labile at ambient temperature.

II. CHLOROPHENOLIC POLLUTANTS AND RELATED SUBSTANCES

The presence of the electron-donating hydroxyl group leads to a very rapid reaction of the monochlorophenols; this is consistent with the view (Ayres and Gopalan, 1976) of the reaction mechanism as one of electron donation by the substrate to ruthenium tetroxide. It is advantageous to oxidize heavily chlorinated phenols as their phenolate salts to counteract the electronegative substituents.

When the amount of oxidant is limited a number of relatively complex intermediates have been isolated (Ayres et al., to be published).

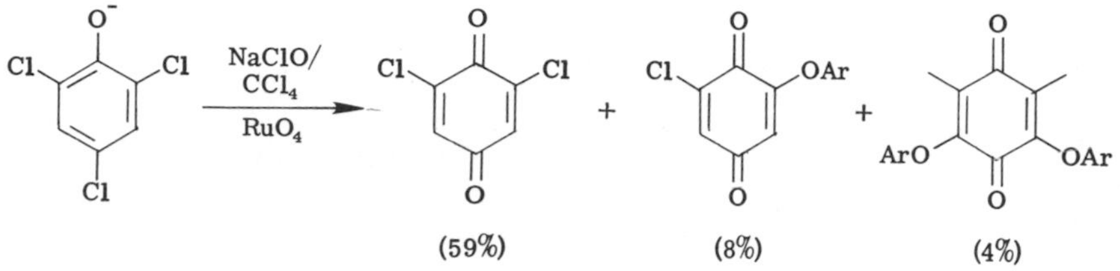

$NaClO/CCl_4$, RuO_4

(59%) (8%) (4%)

Pentachlorophenol affords the coupled product.

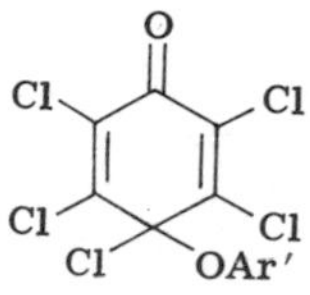

These products are also obtained by the action of other oxidizing agents which act through radical intermediates, for example, nitric acid and benzoyl peroxide. Provided a sufficient amount of the secondary oxidant is available, ruthenium tetroxide completely degrades these intermediates: it will, in fact, oxidize tetrachlorobenzoquinone (Ayres and Levey, to be published).

III. MECHANISM OF THE REACTION

The 2,5- and 2,6-dichlorophenoxy radicals have been characterized (Ayres and Gopalan, 1976) by esr following treatment of the parent chlorophenols with ruthenium tetroxide in aqueous acetone.

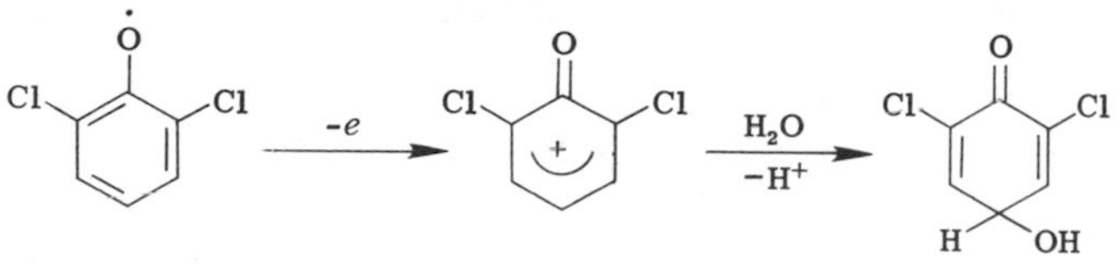

$-e$; H_2O, $-H^+$

Also relevant to the problems of pollution control was the characterization (Ayres and Gopalan, 1976) of the purple radical cation from oxidation of 2-(4-methylphenyl)furan in carbon tetrachloride.

Chloro substituents can be introduced

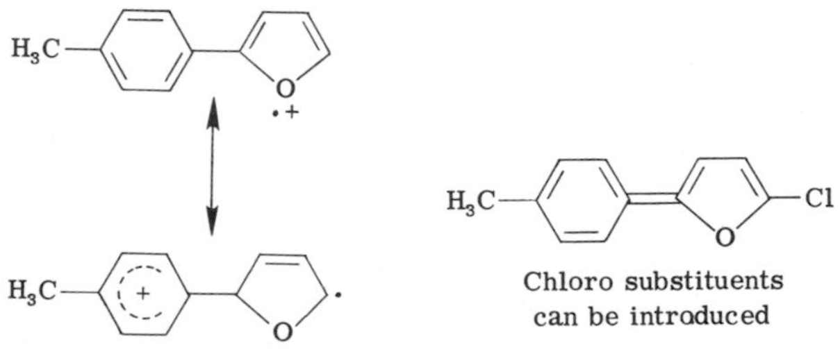

A minor path in this reaction is the capture of chlorine atoms to give the 2-chloro- and 2,4-dichloro-5-(4-methylphenyl)furans. This insertion has been seen with benzene derivatives and results from the generation of chlorine from a ruthenium trichloride precursor: it is only significant when the trichloride is used in stoichiometric quantities.

It is anticipated that chlorodibenzofurans will undergo oxidative degradation in this way. It is also noteworthy that the cation radicals (Baciu et al., 1976; Shine and Shade, to be published) of dibenzo-*p*-dioxins are readily generated by electrolysis.

IV. OXIDATIVE CONTROL

In our laboratory work on the dioxins themselves was begun (Ayres, 1981) with a study of the degradation of 2,7-dichlorodibenzo-*p*-dioxin (0.28 μmol ml^{-1}) in carbon tetrachloride (70 μg ml^{-1}) by ruthenium tetroxide (x4.10^3 μg ml^{-1}, 24 μmol ml^{-1}) in sufficient excess (ca 8 equiv.) to ensure first-order kinetics. A study of the reaction rate at three different temperatures showed that the rate increase for a 10°C temperature rise was about 2.4-fold: the half-life of DCDD at 20°C was about 500 minutes. The progress of the oxidation was followed by quenching the remaining oxidant by the addition of an aliquot of benzene and determination of the residual material by a GLC comparison with a standard solution of DCDD using a 2 m 3% SE30 column at 190°.

DCDD is a useful mimic for TCDD in that the rates of their oxidation by ruthenium tetroxide and also of their photolysis are similar. At 20°C TCDD has a half-life of about 560 minutes in contact with the tetroxide; in refluxing carbon tetrachloride the lifetime is greatly reduced.

TABLE I. Products of the Degradation of Some Aromatic Compounds by Ruthenium Tetroxide

Substrate	*Equiv. of chloride ion (as AgCl)*	*Equiv. of carbon dioxide (as $BaCO_3$)*
Benzene	NA[a]	4.06 (67%)
Phenol	NA	5.03 (84%)
2-Chlorophenol	1.00 (100%)	--
4-Chlorophenol	--	2.29 (38%)[b] 5.39 (90%)
2,4,5-Trichlorophenol	2.64 (88%)	--
Pentachlorophenol (as alkali salt)	4.03 (81%)	5.33 (89%)
Chlorobenzene	0.76 (76%)	--

[a] *NA, not applicable.*

[b] *Result corresponding to the faster initial oxidation to oxalic acid.*

The extension of the work to OCDD is relevant because this substance occurs (Plimmer, 1973) in commercial samples of pentachlorophenol (PCP), widely used for timber treatment. Degradation of OCDD to dioxins of lower chlorine number proceeds by photolysis (Dobbs and Grant, 1979; Nestrick et al., 1980). In solution 2,3,7,8-TCDD is rapidly photolyzed relative to higher homologs, but its absolute reaction rate is very much slower as a dispersion on glass, and its relative rate is slower than most higher homologs. Photolysis of PCP residues may, therefore, produce significant amounts of 2,3,7,8-TCDD.

The formation of TCDD *in vivo* is also possible since hexachlorobenzene, a highly resistant substance, has been shown to undergo reductive metabolism in rats (Koss et al., 1976). The dechlorination of OCDD by soil organisms is also possible.

OCDD is destroyed by ruthenium tetroxide in excess in refluxing carbon tetrachloride; the half-life of the reaction is about 15 minutes.

Table I (Ayres and Levy, in press) shows that destruction of the aromatic ring is essentially complete, which is consistent with the dearth of information about fragmentary products in the existing literature. The chlorophenols themselves are ultimately degraded to chloride ions and carbon dioxide:

$$C_6H_4{\cdot}Cl{\cdot}OH + 11[O] \rightarrow 2(CO_2H)_2 + 2CO_2 + HCl$$

then

$$2(CO_2H)_2 + 2[O] \rightarrow 4CO_2 + 2H_2O$$

and also

$$C_6Cl_5{\cdot}OH + 11[O] \rightarrow 6CO_2 + 5HCl$$

Although no direct evidence of PCDD oxidation products is available, the complete destruction of PCP in 18 hours in water at 30°C suggests that organochlorine fragments do not survive.

REFERENCES

Ayres, D. C. (1981). *Nature (London) 290,* 323.
Ayres, D. C. and Gopalan, R. (1976). *J. Chem. Soc. Chem. Commun.,* p. 890.
Ayres, D. C. and Levy, D. P., to be published.
Ayres, D. C., and Levy, D. P., in press.
Ayres, D. C., Gopalan, R., and Levy, D. P., to be published.

Baciu, I., Hillebrand, M., Sahini, V. E., and Volanschi, E. (1976). *Rev. Roumaine Chim. 21*, 485.
Dobbs, A. J., and Grant, C. (1979). *Nature (London) 278*, 163. Nestrick, T. J., Lamparski, L. L., and Townsend, D. I. (1980). *Anal. Chem. 52*, 1865. Shine, H. J., and Shade, L. R., to be published.
Koss, G., Koransky, W., and Steinbach, K. (1976). *Arch. Toxicol. 35*, 107.
Lee, D. G., and van den Engh, M. (1973). *In* "Oxidation in Organic Chemistry" (W. S. Trahanovsky, ed.), Vol. VB, p. 177. Academic Press, New York.
Plimmer, J. (1973). *J. Agr. Food Chem. 21*, 90.

DISCUSSION

DR. KOLBYE: Would bromine substitution for chlorine in these compounds give rise to a final product more susceptible to photodegradation?

DR. AYRES: We do not know if this would be the case.

DR. KORTE: I was very much interested in your data on pentachlorophenol, its metabolism and so forth. We have studied its oxidation for 5 years under so-called constant conditions. The best destruction was only 10% on the basis of studies of labeled compounds. Does this compare to your data?

DR. AYRES: The cost of ruthenium tetroxide and the risk of toxicity of this substance is a possible problem in the use of RuO_4 as an oxidizing agent. The toxicity is a risk and compares with ozone. Used in closed systems it poses no hazard and has not been a problem for laboratory workers. However, it would not be suitable for use in soils.

DR. SHEPARD: Dr. Crosby's paper on photochemical degradation of dioxins will be discussed next.

ISBN 0-12-193160-9

CHAPTER 7

METHODS OF PHOTOCHEMICAL DEGRADATION OF HALOGENATED DIOXINS IN VIEW OF ENVIRONMENTAL RECLAMATION

Donald G. Crosby

Department of Environmental Toxicology
University of California
Davis, California

I. INTRODUCTION

Polychlorinated dioxins have generally been considered to be unreactive. This observation led to early speculation that highly toxic isomers such as TCDD (2,3,7,8-tetrachlorodibenzo-*p*-dioxin) might be virtually undegradable. However, TCDD was found to be rapidly degraded in alcoholic solutions by ultraviolet (uv) light (Crosby et al., 1971), with the formation of successively less chlorinated dioxins until dechlorination was complete. The dechlorination was much slower in thin films of neat TCDD and appeared not to occur in aqueous TCDD solutions (Plimmer et al., 1973), although it took place readily in a wide variety of organic solvents (Liberti et al., 1978).

The purpose of this chapter is to review the use of photochemical methods for degradation of halogenated dioxins and to show how our simple initial observation has been extended toward the promise of environmental reclamation of dioxin-contaminated areas.

ISBN 0-12-193160-9

II. THEORETICAL BACKGROUND

The chemical basis for the photolytic degradation of halogenated dioxins and related compounds has been presented by Crosby (1978). Three rather obvious requirements must be met if degradation is to proceed: light energy must be absorbed by the compound, light of the appropriate wavelengths must be available at meaningful intensities, and, as photochemical degradation involves reductive dechlorination, a source of excess reducing (hydrogenating) power must be present.

The chlorinated dioxins absorb uv light with maxima in the range of 290 to 320 nm; TCDD absorbs maximally at 307 nm with a bathochromic tail extending to above 350 nm (Fig. 1)(Crosby et al., 1973). Absorption coefficients are relatively high (TCDD has $\varepsilon = 6030$ at 307 nm). Energy absorption in that spectral region is sufficient to break aromatic C-Cl bonds to form radicals which undergo photoreduction by abstraction of hydrogen atoms from C-H compounds in the surrounding medium or from other H donors (Crosby, 1975); alcohols, petroleum hydrocarbons, liquid herbicide formulations, and natural oils have all been used, but TCDD itself apparently is not a suitable donor (Crosby et al., 1971).

The ensuing degradation reactions are shown in Fig. 2 for TCDD. With each successive loss of chlorine, the reaction rate increases, while toxicity of the resulting product diminishes (Crosby et al., 1971; Esposito et al., 1980). At some point, the dioxin ring system is also cleaved to give phenoxyphenols, which, in turn, are further degraded (Plimmer et al., 1973). The expelled chlorine atoms appear as HCl or combined with the H donor in place of the original hydrogen (Crosby and Hamadmad, 1971). Chlorinated dibenzofurans behave similarly (Crosby et al., 1973; Hutzinger et al., 1973).

For environmental reclamation, the third factor -- a source of sufficiently intense uv light -- arises as the principal technical problem. The low wavelength cutoff of sunlight normally lies a little below 300 nm (Koller, 1965), allowing absorption by the halogenated dioxins, but intensity is low, below about 330 nm (Fig. 3). Although direct sunlight is not necessary, as a significant proportion of solar uv is reflected from open sky (Koller, 1965), the period of maximum intensity is both seasonally and diurnally limited. A number of artificial sources of uv radiation are commercially available, and each has its advantages and disadvantages.

It is important, too, that appropriate wavelengths of uv radiation not be absorbed by the surrounding medium before they can reach the dioxin. For example, soil particles effectively block penetration of all uv wavelengths; benzene and other aromatic solvents absorb the 254 nm emission of the

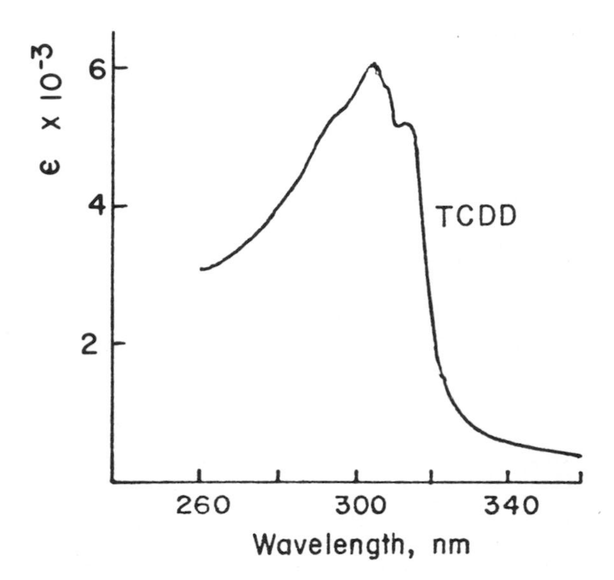

Fig. 1. Ultraviolet absorption spectrum of TCDD (in methanol).

mercury arc, and strongly colored impurities can greatly reduce the penetration of uv radiation into a dioxin solution.

IV. EXAMPLES OF ENVIRONMENTAL RECLAMATION EXPERIMENTS

Where the three essential conditions can be met, sunlight provides an adequate and most convenient source of photochemical energy for dioxin photodegradation. The TCDD impurity in commercial phenoxyester herbicide formulations, as well as in the chemical warfare defoliant known as "Agent Orange" (mixed butyl esters of 2,4-D and 2,4,5-T), was decomposed within a few hours in June sunlight (Fig. 4) on nonporous surfaces or plant leaves(Crosby and Wong, 1977; Wong and Crosby, 1978). The liquid esters quickly were absorbed into soil, and only that portion remaining on the surface and exposed to light was affected. The esters served as the H donors.

In a subsequent field experiment, a commercial 2,4-D/2,4,5-T ester formulation, shown to contain about 0.02 ppm of TCDD, was applied by helicopter for brush control in a northern California experimental forest area in October. Inert panels placed horizontally at ground level throughout the treated area received the herbicide spray and its contained dioxin as measured by extraction and analysis, but the TCDD level declined by more than one-half within the first day after application and was reduced to undetectable after the

second day (Crosby),

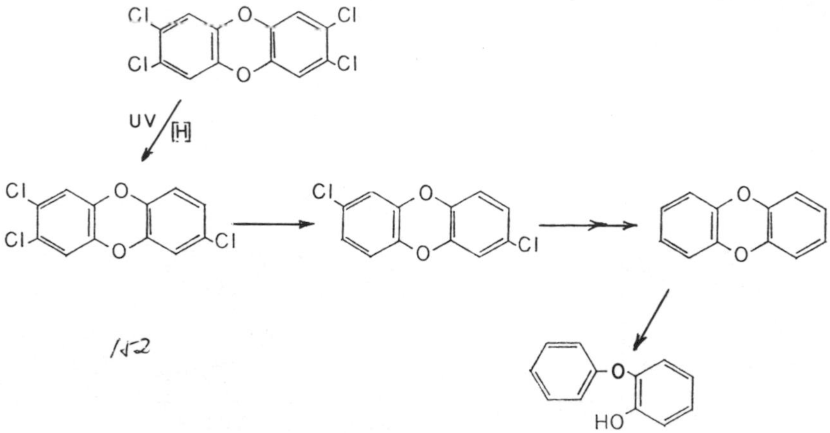

Fig. 2. Photochemical degradation route of TCDD in the presence of a hydrogen-donating solvent. See Plimmer et al. (1973).

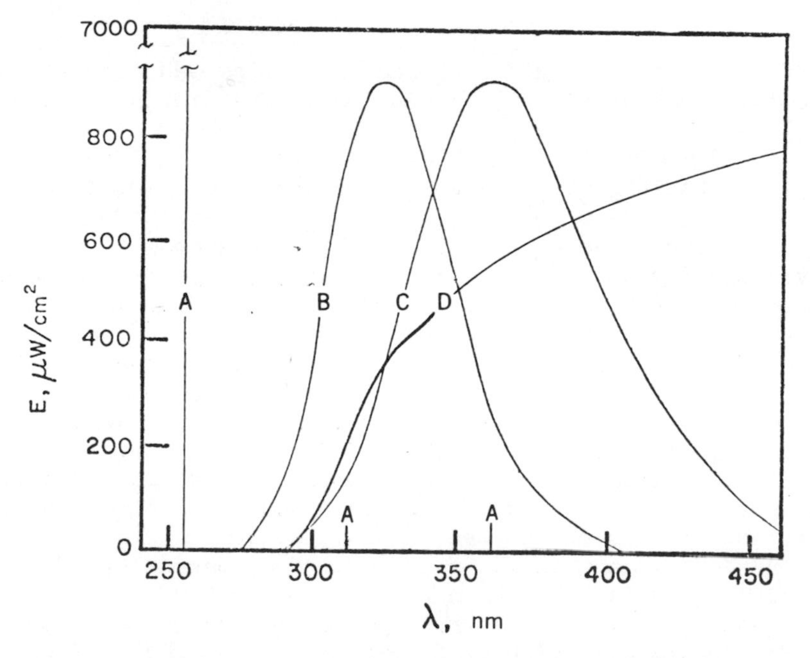

Fig. 3. Spectral energy distribution of common uv light sources including 34 W low-pressure, mercury arc (A), 40 W fluorescent mercury arcs (B,C), and summer sunlight (D). Unshielded Xe arc provides all wavelengths 250-450 nm. See Koller (1965).

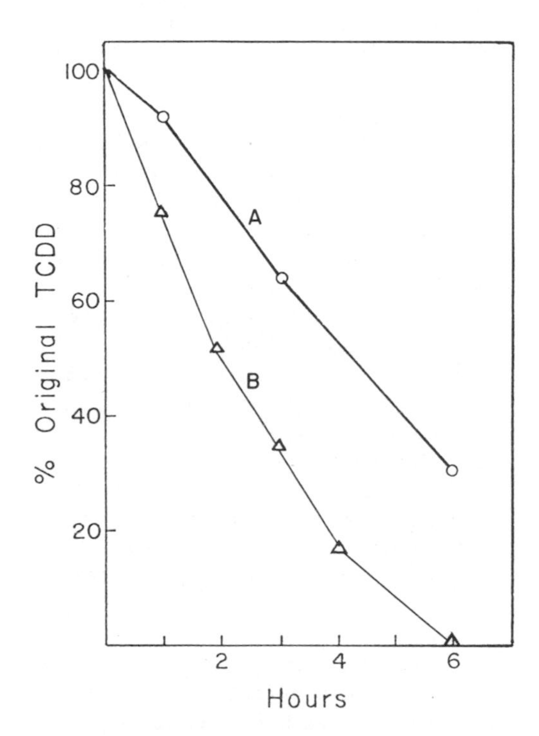

Fig. 4. Sunlight photolysis rate of Esteron 2,4,5-T formulation on a glass plate (A) and Agent Orange on a leaf surface (B). Adapted from Crosby and Wong (1977).

second day (Crosby, unpublished, 1978). We interpret these results to signify that in instances where dioxins occur as contaminants dissolved in an organic medium (such as a pesticide formulation), photochemical degradation in the environment often will be rapid and spontaneous.

They also indicate that similar results might be expected if a TCDD deposit could be brought into solution in a hydrogen-donating solvent and irradiated. This was successfully demonstrated during attempts to deal with environmental contamination by TCDD following a much-publicized industrial accident at Seveso, Italy, in 1976. TCDD levels averaging 26.2 $\mu g/m^2$ on grass were reduced by over 80%, proportionate to the available uv energy, within 9 days following the spraying of contaminated test plots with an aqueous emulsion of olive oil to act as H donor (Crosby, 1978, Wipf et al., 1978). The technique eventually could have been successful in removing or lowering much of the dioxin contamination in the Seveso area, but the opportunity was lost to approaching winter when another of the three necessary ingredients -- sunlight -- became unavailable.

From the uv absorption spectrum (see Fig. 1), it is apparent that TCDD will absorb the intense 254 nm light radiated by commercial mercury-arc lamps, although not optimally. A process, never reported in detail (Esposito et al., 1980), was designed by a herbicide manufacturer in the early 1970's to recover and purify the millions of pounds of valuable 2,4-D and 2,4,5-T represented by unused Agent Orange being held in storage by the United States Government. These mixed butyl esters were to be base-hydrolyzed, the neutral impurities including TCDD solvent-extracted out of the hydrolyzate, and the organic extract irradiated as a thin, flowing film by mercury-arc lamps. The extraction solvent, such as butyl alcohol, was to be the H donor. While the procedure was successful in removing and destroying TCDD on a laboratory scale, economics and government regulations never permitted its practical use.

Recently, Kearney et al. (1981) reported experiments on the use of mercury-arc lamps to degrade TCDD (and 2,4,5-T) in aqueous wastes resulting from spray application of the herbicide. In this case, the H donor again was the 2,4,5-T ester and/or its organic solvent carrier. Excess spray mix, unused concentrate, or spray-tank washings could be collected, recirculated through a uv-irradiated reactor, and the partially degraded mixture then released to soil for natural microbiological destruction. The initial results appeared very promising, and experimental work is continuing at the present time. Wong and Orbanosky (1979) and Cavolloni and Zecca (1977) separately reported the bench-scale destruction of TCDD in water by a combination of uv light and ozone. While these results also appeared promising, it seems unlikely that photoreduction is involved in this instance, and the chemical mechanism remains unexplained.

The practical utility of photochemical removal of TCDD from chemical waste has been demonstrated only recently (Exner et al., 1981). Nearly 4300 gallons of reactor wastes containing 343 ppm of TCDD were found in a storage tank at a chemical factory in Verona, Missouri -- a total of about 13 pounds of the dioxin. The decision was made to attempt to detoxify the TCDD by photochemical reduction, and it was found to be extracted away from the aqueous sludge satisfactorily with hexane, which also served as H donor. The extract subsequently was circulated through a series of glass-lined reaction chambers equipped with industrial-scale mercury-arc lamps, isopropyl alcohol being added to eliminate fouling of the lamps by uv-generated polymeric film. Eventually, the dioxin content of the waste was lowered by over 99.9%, to less than 0.3 ppm, with essentially complete photochemical dechlorination of the TCDD as well as extensive destruction of several other chlorinated waste products.

IV. DISCUSSION

Even these few existing examples demonstrate conclusively the promise of photochemical degradation of halogenated dioxins and related compounds in view of environmental reclamation. However, it is worthwhile to give a few words of technical comparison of uv sources, H donors, and the advantages and limitations of photochemical methods.

A. *UV SOURCES*

Where it can be used, sunlight seems to be the obvious choice of uv source -- cost-free, safe, and effective against TCDD. It also can lead to obvious difficulties -- maximally effective for only limited periods, dependent largely on fair weather (although a proportion of this uv can penetrate clouds), and, so far, of limited adaptability to factory-type operation. On the other hand, mercury-arc lamps, especially in tubular configurations, are readily adapted to plant processing and are available in a wide range of sizes and intensities. However, over 95% of the output of a low-pressure arc is at 254 nm, where TCDD has an uv absorption minimum, while that of a medium-pressure arc is mostly above 360 nm. Energy efficiency generally is below 20%, and the high voltage requirements, short wavelength, and heat generation can present serious health hazards. While the fluorescent arc-lamps are cooler, safer, and more efficient, the available intensities remain low. Xenon arcs have many disadvantages, not the least of which is high cost, but they do produce uv energy at high intensity over all of the dioxin absorption range.

B. *H DONORS*

Theoretically, almost any organic liquid having a large proportion of hydrogen and not highly uv-absorbing in the same wavelength range as the dioxin would suffice. However, several practical factors limit the number of possibilities. Often it will be undesirable for the H donor to have appreciable water solubility, while not as important in a closed system, an H donor for outdoor use must have low volatility (a high boiling point); toxicity and flammability always are important considerations; and cost frequently would be a limiting factor. Among water-soluble donors, glucols and glycol ethers such as Carbitols and Cellosolves meet the requirements. Many natural vegetable and mineral oils would be useful as water-insoluble donors, as would petroleum distillates, except for flammability. The dioxin-containing pesticide formulations generally serve as their own H donors.

C. ADVANTAGES AND LIMITATIONS

Photochemical methods of degradation usually can be relatively simple, adaptable to large-scale operation (with some increase in technical complexity), versatile, and often comparatively inexpensive. However, the greatest advantage -- one which can hardly be stressed too much -- is that the dioxins are *permanently destroyed*. While removal of TCDD from contaminated pesticides by adsorption on carbon, or excavation of contaminated soil, only serve to move the contaminant from one place to another, photolysis (like combustion) detoxifies it forever.

However, there are many practical limitations. Photodegradation generally would be predicted to be unsatisfactory for direct decontamination of soil or other dense particulate matter which the radiation cannot penetrate, although Bertoni et al. (1978) reported uv degradation of soil-borne TCDD in the presence of added H donor. u.c. irradiation cannot be energy efficient -- the quantum yield, or number of molecules reacted per photon absorbed, is only 0.009 -- although this probably would be unimportant for most environmental photolysis in sunlight. Further, the intense uv radiation from a mercury or xenon arc, the high voltages required, and any flammable solvents are dangerous. Other problems and costs are discussed by Plimmer (1978) and by Bulla and Edgerley (1968).

Despite the difficulties, photodegradation seems to have many potential applications (Esposito et al., 1980). For example, TCDD on interior surfaces in a room can be destroyed by irradiation with a portable uv lamp (Bertoni et al., 1978), or the surfaces can be washed with detergent and the washings irradiated to destroy the dioxin (Botre et al., 1979). TCDD residues in hard-to-reach locations might be destroyed with a uv laser beam. As TCDD is steam-volatile, contaminated surfaces could be steam-cleaned and the condensate collected and irradiated after addition of H donor; contaminated soil could be treated with a nonvolatile H donor immediately after tillage and sunlight photolysis allowed to take place before retillage; contaminated paper or cloth could be rinsed with a glycol or detergent solution before normal incineration, with subsequent irradiation of the washings. Such methods are simple and inexpensive compared to the present alternatives.

Certainly, photochemical treatment will not be feasible in all possible situations, but *it works*. I propose that it be tried on a modest scale before other, more heroic, destructive, and costly measures are turned to.

REFERENCES

Bertoni, G., Brocco, D., DiPalo, V., Liberti, A., Possanzini, M., and Bruner, F. (1978). Gas chromatographic determination of 2,3,7,8-tetrachlorodibenzodioxin in the experimental decontamination of Seveso soil by ultraviolet radiation. *Anal. Chem. 50*, 732.

Botre, C., Memoli, A., and Alhaique, F. (1979). On the degradation of 2,3,7,8-tetrachlorodibenzoparadioxin (TCDD) by means of a new class of chloroiodides. *Environ. Sci. Technol. 13*, 228.

Bulla, C. D. and Edgerley, E., Jr. (1968). Photochemical degradation of refractory organic compounds. *J. Water Pollut. Control Fed. 40*, 546.

Cavolloni, L. and Zecca, L. (1977). Decomposition of TCDD by ozone. *Med. Term. Climatolog. 34*, 73.

Crosby, D. G. (1975). Herbicide photodecomposition. *In* "Herbicides: Chemistry, Degradation, and Mode of Action" (P. C. Kearney and D. D. Kaufman, eds.). p. 835. Dekker, New York.

Crosby, D. G. (1978). Conquering the monster: the photochemical destruction of chlorinated dioxins. *ACS Symp. Ser. 73*, 1.

Crosby, D. G. and Hamadmad, N. (1971). The photoreduction of pentachlorobenzenes. *J. Agr. Food Chem. 19*, 11771.

Crosby, D. G. and Wong, A. S. (1977). Environmental degradation of 2,3,7,8-tetrachlorodibenzo-*p*-dioxin (TCDD). *Science 195*, 1337.

Crosby, D. G., Wong, A. S., Plimmer, J. R., and Woolson, E. A. (1971). Photodecomposition of chlorinated dibenzo-*p*-dioxins. *Science 173*, 748.

Crosby, D. G., Moilanen, K. W., and Wong, A. S. (1973). Environmental generation and degradation of dibenzodioxins and dibenzofurans. *Environ. Health Perspect. 5*, 259.

Esposito, M. P., Tiernan, T. D., and Dryden, F. E. (1980). Dioxins, Environ. Protect. Agency Publ. EPA-600/2-80-197, p. 257. Cincinnati, Ohio.

Exner, J. H., Johnson, J. C., Ivins, O. D., Wass, M. N., and Miller, R. A. (1981). A process for destroying tetrachlorodibenzo-*p*-dioxin in a hazardous waste. *Abstr. 182nd Nat. Meet. ACS*, New York.

Hutzinger, O., Safe, S., Wentzell, B. R., and Zitko, V. (1973). Photochemical degradation of di- and octachlorodibenzofuran. *Environ. Health Perspect. 5,* 267.

Kearney, P. C., Plimmer, J. R., and Li, Z. M. (1981). Solution photolysis as a pretreatment in soil disposal of chlorinated organic wastes. *Abstr. 180th Nat. Meet. ACS,* Atlanta, GA, PEST 48.

Koller, L. R. (1965). "Ultraviolet Radiation," 2nd Ed. Wiley, New York.

Liberti, A., Brocco, D., Allegrini, I., and Bertoni, G. (1978). Field photodegradation of TCDD by ultra-violet radiations. *In* "Dioxin: Toxicological and Chemical Agents" (F. Cattabeni, A. Cavallaro, and G. Galli, eds.), p. 195. SP Medical and Scientific Books, New York.

Plimmer, J. R. (1978). Approaches to decontamination or disposal of pesticides: photodecomposition. *ACS Symp. Ser. 73,* 13.

Plimmer, J. R., Klingebiel, U. I., Crosby, D. G., and Wong, A. S. (1973). Photochemistry of dibenzo-*p*-dioxins. *Adv. Chem. Ser. 120,* 44.

Wipf, H., Homberger, E., Neuner, N., and Schenker, F. (1978). Field trials on photodegradation of TCDD on vegetation after spraying with vegetable oil. *In* "Dioxin: Toxicological and Chemical Aspects" (F. Cattabeni, A. Cavallaro, and G. Galli, eds.), p. 201. SP Medical and Scientific Books, New York.

Wong, A. S. and Crosby, D. G. (1978). Decontamination of 2,3,7,8-tetrachlorodibenzodioxin (TCDD) by photochemical action. *In* "Dioxin: Toxicological and Chemical Aspects" (F. Cattabeni, A. Cavallaro, and G. Galli, eds.), p. 185. SP Medical and Scientific Books, New York.

Wong, A. and Orbanosky, M. (1979). Ozonation of 2,3,7,8-tetrachlorodibenzo-*p*-dioxin (TCDD) in water. *Abstr. 178th Nat. Meet. ACS,* Washington, D. C., PEST 83.

DISCUSSION

DR. SHEPARD: This presentation was fascinating, and I hope, Dr. Crosby, your theory is accurate, for we have been telling veterans that probably their exposure was limited by virtue of the fact that photodegradation was taking place. Another ingenious idea was proposed, namely, the use of photodegradation to analyze for TCDD.

DR. BLAIR: 2,4- and 2,4,5-T, contaminated with TCDD, has been in use for over 30 years. In some cases, this was along the right-of-way of railroads, where if there was a build-up of dioxins resistant to degradation, we should have information on the presence of TCDD through United States food basket studies -- TCDD would have been found in food. In fact, it realy does not seem to be in the environment at all. This supports the findings in Dr. Crosby's studies.

DR. COULSTON: Dr. Crosby's studies support the kind of work Professor Korte has done in Europe on photodegradation. Other groups have reached similar conclusions. The point was made that if TCDD is sprayed from the air, 24 hours later it can no longer be found on the ground -- dioxin is gone! It must be remembered that only 0.05 ppm or less dioxin is found in pure 2,4,5-T. Many years ago, the manufacturers agreed to keep the concentration of dioxins below 0.05 ppm. Actually, in practice, it is less than 0.01 ppm. In a recent meeting it was noted that women were exposed to dioxin in 2,4,5-T. It was stated that when this was diluted to prepare the spray for dissemination by airplane, the dioxin content could not be measured. For practical purposes, it was

ISBN 0-12-193160-9

nonexistent. It is this point that must be understood. It has been emphasized that even the "flagmen," who were soaked with the spray, had no measurable dioxin on their clothing or anywhere else. That the women in the Alsea Basin were overexposed to TCDD is somewhat suspect from a technical point of view. Second, the data do not lend themselves to any kind of positive statistical analysis that dioxins in the 2,4,5-T caused any miscarriages. This was the conclusion reached and these data have been published.

DR. CROSBY: It should be emphasized that in our work the commercial form of 2,4,5-T Esceron and not a "spiked" spray with added TCDD was used. It turned out that Esceron contained only 18 ppb of TCDD. It is hard to imagine how a spray over 50 acres containing such a diluted material could even be detected. If it was also affected by sunlight for a few hours. Thus, it is no wonder that TCDD was undetectable.

DR. POCCHIARI: No one doubts the occurrence of photodegradation. Actually, after 1 month we found that more than 50% of TCDD had disappeared. Our experiments have also shown that sunlight does not penetrate a few millimeters into the soil and, of course, there is no photodegradation. Rain will also cause the agent to penetrate the soil making it hard to remove. This is a physical action. In large areas requiring decontamination, it is very important to rapidly clean the soil so that these several factors do not prevent effective measures.

DR. KORTE: It is very important to point out that chemicals must be degraded and that this has been going on for years. Diffused daylight and sunlight must be very effective in this process or else the entire world would be covered -- there is no other explanation. However, we should do more work on soil, because we do not know how to measure the volatility from soil. More investigations are needed of soil chemistry to provide the answers to all of these questions.

DR. BARNES: Do you know what fraction, if any, is due to volatility?

DR. CROSBY: In our experiments, no. In the references mentioned, determinations could be made because controls could be carried out in the dark under the same conditions. Part of the loss was due to volatilization. TCDD can be steam distilled and can be volatilized by itself. Perhaps high-pressure steam could be used for cleaning buildings and perhaps even soil. However, it is necessary to have a hydrogen donor. This is very important if TCDD is to be completely removed.

CHAPTER 8

EXPERTS, AUTHORITIES, AND PROPHETS*

Etcyl H. Blair

Vice President
Director, Health and
Environmental Sciences
The Dow Chemical Company
Midland, Michigan

Good evening. I am very honored to be here tonight to address such an elite group. The implication is that I am somehow member of such an elite. I will confess to having made some minor contributions to the literature on dioxins, but what I am going to say to you tonight will have very little to do directly with dioxins.

It is not that I have any doubts about the important nature of this conference, or that I particularly want to provide any sort of diversion from the weighty and complex considerations you have before you. Instead, I want to discuss a phenomenon which I think is a critical part of the backdrop against which scientific meetings such as this one are held.

The phenomenon is that we are surrounded in science today by a host of conflicting truths and an army of experts standing like sentinels on a beach, ready to charge forward at a moment's

**Address presented at the Banquet Dinner, October 5, 1981.*

ISBN 0-12-193160-9

notice to support an embattled truth. We have your truth, and my truth, and his truth, and maybe several other truths. And it seems that every truth has at its beck and call an expert who is ready to recite the data and show us why that truth is *the* truth.

I do not say this lightly, nor in an effort to disparage someone else who has a version of the truth that differs from mine, although those of you who know me realize that I am ready to do battle with those for whom the truth is set upon rubber pillars in a field of shifting sand. I say it because I recognize that truth becomes dogma when the experts are called in, and I think all of us as scientists should be on guard against that event. We all want truth on our side, but there are--sadly--those who do not genuinely want to be on the side of truth.

Experts are necessary people, but they are no more infallible than others and, indeed, the history of the world--not just the scientific world but the world at large--is full of people who were sober, qualified, individuals--well-trained, experienced, capable and even eminent in their fields. They may have every right to the respect and honor which history has finally accorded them, but somewhere along the way a dogma reared its ugly head and these respected people simply did not deserve that respect on a given subject, or for a given time frame, or in hindsight--whatever.

I say this even at the risk of indicting myself with my own words. I mean that I am not infallible, and there is every risk in what I am saying that years from now some of you will stand at a podium somewhere and recount the story of the scientist who warned of dogmatic experts and, in the final analysis, turned out to be one himself.

I would call your attention, for example, to the remarkable Danish nobleman, Tycho Brahe. He was born in 1546, and when he was only 30 the king of Denmark made him landlord of the island of Ven and gave him enough money to build a house and astronomical observatory there. Brahe became the greatest observational astronomer the Christian world had known. He used only a giant quadrant on his lonely island base to make a 20-year study of the skies and produced a magnificent and comprehensive star catalogue, including tables that gave thousands of accurate "fixes" on the changing positions of the planets against their background of stars.

There was one serious drawback. Despite the fact he had access to the work of Copernicus, and that in his later years he worked with the astronomical genius, Johannes Kepler, what Brahe was trying to do was prove that Ptolemy was right--that the sun revolved around the earth. And he died convinced that it did. Had there been congressional committees at the time or need for an expert's opinion about the earth's relationship to the sun, or vice versa, Brahe almost certainly would have been called. And who knows whether it might have been against the law to launch an Apollo satellite toward the moon?

Take another example. The eminent German biochemist and Nobel Prize winner, Otto Warburg, is highly respected among us for his pioneering work in cellular respiration and photosynthesis. He was widely recognized and respected and his word carried great weight in the 1920's and the 1930's. Warburg had an Achilles heel, however, a theoretical misconception he carried for more than 50 years, which is the approximate length of time he maintained that the primary cause of cancer was, of all things, a defect in the respiratory system of the victims of the disease. Had his era seen the passionate, obsessive interest in cancer that we live with today, there is little question that Warburg's belief would have colored a great deal of the work on breaking the many codes of the mysterious family of diseases we categorize as cancer. And who knows how many research dollars were mischanneled into trying to confirm this view of Warburg's?

Well into this century we have examples of giants of science, experts in their fields, who have harbored some rather questionable beliefs. Not to harp on astronomers who can perhaps be pardoned a bit more hypothesizing than the rest of us, but the brilliant American astronomer, Percival Lowell, was one of more than a handful of scientists who convinced themselves that Mars was inhabited. The canals, he became certain, were artificial and he came to this conclusion:

> "Mars is inhabited by beings of some kind or other; this is as certain as it is uncertain what those beings may be."

Lowell would certainly not have opposed space probes of Mars to learn more about the planet, but suppose his conviction had concerned some more earthbound subject. Do you not think that such a man might have convinced others that no further evidence was necessary and that legislation was needed to control a postulated truth?

No one is really certain how the universe began, but we have no lack of theories. Again, they may be pardoned some excesses of hypothesis, but we have many, many experts from whom to choose. The Belgian, Georges Lemaitre, was an early exponent of "The Big Bang" theory, about 1930. "About 10 million years ago", he said, "the universe was contained in a primal atom"--a superdense "cosmic egg" which exploded.

Thirty-five years later, the American astronomer, Allan Sandage, adopted this theory but expanded it considerably.

His "Pulsating Universe" theory is that the universe is created, destroyed and recreated in 80-billion-year cycles.

Not so, say the British cosmologists Bondi, Gold and Hoyle --their "Steady State" theory is that the universe has always existed and that matter is created, apparently out of nothing, at the rate of 62 atoms of hydrogen per cubic inch every billion years.

Every one of them an expert, but thank God you and I do not have to base daily decisions affecting our careers or the financial resources of our institutions on choices between such alternatives.

Science is not the only field afflicted by expert opinion on which such misleading convictions are held.

George Osborne was a distinguished professor of law at the prestigious Stanford University Law School in 1952. He called that year's graduating class the "dumbest" he had ever taught. Were he alive today, his expert judgment would have fallen into disrepute; the class included two of the nine justices of the Supreme Court of the United States, William Rehnquist and Sandra Day O'Connor; the governor of Utah; five current members of the California superior court; 1960's activist and "hippie" lawyer Jerry Rosen, and Forrest Shumway, chairman of the $4.3 billion conglomerate Signal Companies.

Other "experts" have performed miserably. Using colossally bad judgment, Roebuck sold out to Sears in 1895 for $25,000; today Sears sells goods worth that much in 16 seconds. A highly successful Hollywood producer scrawled a one-sentence rejection note on the manuscript of "Gone With the Wind", and Seward's folly was the purchase of Alaska from Russia for $7,200,000.

Perhaps my point has been made by this time. Experts abound, and there is no lack of market for their opinions. The market is not limited to an uninformed public; there are highly placed people in government, in the professions, in the media, in business and occasionally even science itself, who regularly turn to their favorite experts for confirmation of dubious data.

There is perhaps no field in which this phenomena is more apparent--or more frightening--than in the field of cancer research. I may be in danger of gaining a reputation as something of a crank in this area, since I have spoken many times of my concerns, but I seem to have made relatively little headway so I will risk the reputation once again.

The general public has the idea, in the face of a nearly total absence of data, that there is an epidemic of cancer in the world--and especially in the United States--today. It is understandable, I suppose, that the public should have a serious concern about cancer. But if that is understandable, it is equally as incomprehensible to me that the concern should have grown into such a phobia that concrete evidential data should be ignored. The statistics are clear, it seems to me, that life expectancy is still on the rise, that the incidence of cancer deaths (except for lung cancer) is stable or even declining and that there is simply no evidence that we are experiencing a cancer epidemic of even the most isolated proportions. It also seems to me that this is no reason for us to relax our efforts to identify and understand the mechanisms of cancer. Yet one must come to the reluctant conclusion that there are those who have a big stake in cancer research programs, and it would not do to have the public understand that cancer is not a growing threat.

In my most optimistic moments, I would admit that I see some evidence that this particular tide is turning a little, that the public is slowly coming to comprehend not only what the statistics are saying but that perhaps it has been manipulated. But I am old enough, and maybe cynical enough, to know that if that takes place--if the public sees through the clouds of smoke--another concern is just below the horizon and about to make an ominous appearance.

Do I have such a candidate? Yes.

You are dealing with it here. Birth defects.

There are signs, I believe, that the next subject for scientific controversy is malfunctions in human reproduction. I suppose all of us are aware of the statistics in this field and that they have undergone no material change for a considerable period of time.

The statistics show that serious malformation--and there is by no means unanimity on the definition of that term--occur about 22 times for every 1,000 infants who survive the birth experience. That statistic, of course, *follows* other statistics, which are that perhaps 40 percent of all pregnancies result in fetal wastage--spontaneous abortion, miscarriage or stillbirth. We're not sure what that precise number is because many early wastages go unreported or even unrecognized.

Great strides in medical research have been made in recent years, but there have been no significant changes, either up or down, in that rate of serious malformations-22 per thousand.

More important, those great strides in medical research leave us with an important unknown: the cause, or causes, of fully two-thirds of the reported cases of birth defects are unknown. Unknown!

If medical science can tell only one out of three sets of parents about the mechanical reasons--or possibly chromosomal reasons--for the birth of a child with serious malformation, then that leaves the other two with an anguished question: "WHY?"

And into that void of reason--that inability of medical science to ascribe a cause--steps speculation, and the effort to recreate almost on a minute by minute basis, the long months of pregnancy, to find something, some explanation for this gross error that has been visited on two innocent, healthy specimens of the human species.

There are already those people who have seen in this phenomenon opportunities to further careers in the law, in the public interest arena, in the bureaucracy; I feel quite safe in predicting that this unknown--the 14 or 15 births of a thousand which are inexplicably defective--is the next unknown for which chemical agents are destined to be implicated as the causative factors.

And I am equally as certain that various truths about this phenomena will have no lack of experts to support them.

What's the answer to all this?

Well, I have given the matter a great deal of thought, and I hope I don't disappoint you when I say I have nothing really new to offer.

That is, I think science and scientists have been co-opted into non-scientific arenas--mainly the political arena--and there is really no one to blame except scientists. They have allowed themselves to be exploited because the vogue of the last decade or so has called for all sorts of people to abandon objectivity and enter into various frays as partisans of one sort or another. "Advocacy" teaching became "advocacy" journalism, which called for "advocacy" regulatory officials, who in turn sought out "advocacy" scientists. And suddenly everyone had a passion, a cause, an injustice to right, an imbalance to redress.

So, simply, scientists have to get back to science. They have to remember what scientific methodology is all about--why it is that building hypotheses, setting up experiments, designing all the tests they can think of to disprove their hypotheses, is vital to their calling. They must return to disputing data, not other scientists.

I suppose what pleases me most tonight is the feeling that I am among a group of scientists who have not forgotten that they are scientists, have not forgotten what the mission of science really is. Meetings such as this is what science is all about, I believe, and I will close by commending you for the manner in which you are practicing your calling.

I trust that your voices will be heard and that they will serve the truth.

OPENING REMARKS

DR. COULSTON: The acute effects of an accidental release of dioxins are fairly well recognized, but the long-term health impacts are relatively unknown. There have been many predictions of dire effects, but as yet we have not recognized those effects with many of the chemicals that have been in use for 30 to 40 years.

DR. SILANO: Thank you, Dr. Coulston. First, I would like to welcome all of you to the second day of our meeting on human health aspects to exposure to dioxins and environmental reclamation. We had an excellent discussion of the scientific aspects of these issues yesterday and I would like to add that the social part last night at the magnificent banquet was equally enjoyable. The atmosphere was very warm and friendly and although I have been in the States many times now, very rarely have I had the opportunity to feel so welcome. I am sure this goes a long way to make this a rewarding and successful meeting. Now I would like to introduce Professor Korte who will speak on the "Ecologic Chemistry of Dioxins."

DR. KORTE: The chemistry of the chlorinated dioxins and chlorinated benzofurans are of interest and will be discussed next.

ISBN 0-12-193160-9

CHAPTER 9

ECOLOGIC CHEMISTRY OF DIOXINS AND RELATED COMPOUNDS

F. Korte

Institut für Ökologische Chemie
Neuherberg, München
Germany

Occupational exposure to the dioxins and related compounds have been discussed since the early 1950's. Dioxins have thus had a long history as critical chemicals. Chlorinated dioxins, chlorinated dibenzofurans, and all of the other groups of chemicals being discussed in the context of the dioxin problem are not industrial chemicals in the normal sense; they are not intentionally produced for any technical purpose. However, they serve as a good model for the investigation of by-products or secondary products.

A scheme of how chlorinated dioxins are formed is presented in Fig. 1. It provides as an example of the symmetrical tetrachloro isomer, which is a product of major concern, and shows that a number of possibilities exist that such chemicals may be formed from chlorinated phenols. That this occurs under many conditions for a long time has been known for a long time and has been discussed (Conference on Dibenzodioxins and Dibenzofurans, April, 1973, sponsored by the National Institute of Environmental Health Sciences) and the

ISBN 0-12-193160-9

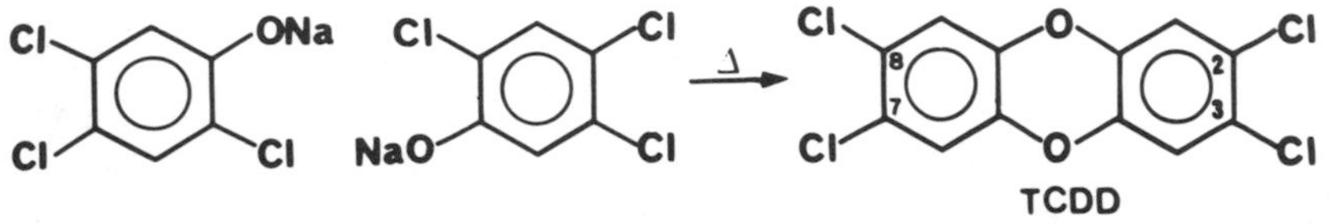

Fig. 1. Formation of chlorinated dioxins.

results published (Jensen and Renberg, 1973). Many isomeric chlorinated dioxins and related chemicals, including polymers, are formed according to the same scheme (Fig. 1).

Dioxins and related compounds occur in different concentrations in a number of technical products, depending on the process used for their preparation. They have thus been unintentionally produced for many years and their toxicity only brought to light by the accidents which occurred. One such case was exemplified by the accident which occurred at Seveso, where significant amounts of toxic chemicals were released into the environment. Another incident occurred in the BASF on November 17, 1953. An uncontrolled reaction during the production of 2,4,5-trichlorophenol in an autoclave (solvent: methanol sodium hydroxide; reaction temperature: 180°C; starting material: 1,2,4,5-tetrachlorobenzene) caused the release of a toxic material which severely affected a total of 42 persons. However, none of the toxic material escaped the building containing the autoclave. Despite extensive measures to decontaminate the building, including painting and silicone treatment, the toxic material remained in the building 5 years after the accident. The BASF finally demolished the entire building in 1968, 15 years after the accident occurred.

The conclusions of a meeting of the European Communities with respect to the decontamination problems for Seveso is presented below, and serves as an example of the magnitude of the entire problem.

> The group was requested to consider decontamination methods and problems, and to indicate suggestions on different

techniques to be applied to selected compartments of the environment like houses, soil, and vegetation.

The group was informed by the regional medical authorities that the level of TCDD accepted outside buildings was 5 $\mu g/m^2$ surface which on average corresponds to 0.01 μg/100 g soil taken at 7 cm depth and 66 cm^2. The practical analytical sensitivity was 0.75 $\mu g/m^2$.

In Region A the average concentration of TCDD was greater than 0.5 mg/m^2; in Region B it was below 3 $\mu g/m^2$. Zone A (approx. 700 inhabitants) was completely evacuated, but not Zone B (approx. 4000 inhabitants). The total amount of TCDD dispersed was on the order of 0.5-3.5 kg and was spread over an area of approx. 100 ha in Region A and approx. 200 ha in Region B.

The group concentrated the discussions on decontamination problems for Zone B since it was considered that recovery of Zone A might either be impossible or a matter of a long period of time.

Houses: Since dioxin was distributed via the gaseous phase, it was necessary to decontaminate the entire house (not only the rooms used by inhabitants, but also uninhabited spaces such as attics, cellars). It was proposed to study the possibility of using vacuum cleaners provided with special dust filter cleaners for the rooms, as well as for the furniture, to paint the walls with protective agents, etc.

Photochemical degradation should be investigated with special reference to mineralizing TCDD to CO_2. This reaction should also be considered in eliminating TCDD from other compartments in the environment. Although in previous cases reported (BASF, Philips Duphar, etc.), destruction of the building was finally necessary, this method should be regarded, in this case, as the last recommended possibility.

Soil: It was generally understood that decontamination should start from lower pollution toward higher pollution areas.

Furthermore, laboratory experiments and *in situ* studies should be proposed for microbiological cultures to be used for degradation (actinomycetes, pseudomonads). Methods should be developed to promote their growth in the field.

Degradation possibilities under the conditions of diffuse and ultraviolet light with possible broad-spectrum sensitizers should be studied and process development work should be started if the laboratory results prove promising.

The applications of heating machines (type used especially for road works) should be considered for application. Digging and incineration of large areas should be considered as the last recommended decontamination possibility.

Vegetation: Since decisions for defoliation have been undertaken and other decontamination measures have been agreed on and practical work has started, no further attention is to be given to this material, but the possibilities of using the local municipal incinerators for the disposal of slightly contaminated plant materials should be investigated.

The following general proposal was agreed upon:

A program of biological monitoring should be maintained by the introduction into the area, if necessary, of domestic and wild species of birds and animals.

Dioxin monitoring should be extended to human tissues where determination of sensitivity might be on the order of parts per thousand.

Even in Zone B, localized regions of contamination which present a potential health risk will undoubtedly remain for a long period of time regardless of decontamination procedures used.

New decontamination procedures should be tested under field conditions on laboratory scale.

In principal there are two different types of processes in which dioxins can be unintentionally formed. One is during technical procedures mainly involving catalytic high-temperature reactions involving chlorinated aromatics, e.g., chlorophenol processes. The other is upon incineration of a number of materials. (Some examples will be given for this later.) Whereas technical processes can be altered to avoid the formation of these chemicals or at least to reduce their concentration to an insignificant amount, this is not possible for incineration processes. Thus, it was possible to limit the dioxin concentrate in 2,4,5-T, which still seems to be an important herbicide, to 0.5 μg/kg. This should be a safe concentration and seems to be acceptable and represents no harm to the environment.

There are, however, a number of impurities which are related to dioxins in technical products.

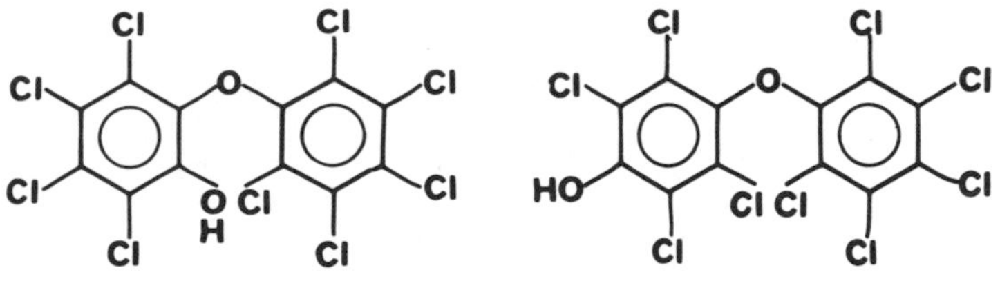

PREDIOXIN ISOPREDIOXIN

Fig. 2. Formula of predioxin and isopredioxin.

Impurities, called predioxins or isopredioxins (Fig. 2), are contained in technical chlorophenols and were known in the early 1970's (Jensen and Renberg, 1973). Chlorodibenzofurans are impurities formed during the production of chlorophenols via diphenylethers and dihydroxichlorodiphenyls, respectively.

More recently, we have investigated the neutral and phenolic impurities in different technical pentachlorophenols. These are given in Figs. 3 and 4. The phenolic components are various chlorophenols and occur in varying amounts in different products. The neutral components in PCP mixtures are chlorodioxins, chlorodibenzofurans, chlorodiphenylethers, and chlorobiphenyls. The concentrations of these impurities vary widely: 0.03-2500 μg/gm for the dioxins and dibenzofurans and 0.1-8% for the chlorophenols.

Although much is known on the formation of dioxins and related compounds in different concentrations in different pyrolytic and incineration processes, work is continuously being done in this area. A few examples are given in Table I. The formation of polychlorinated dibenzodioxins from the pyrolysis of chlorobenzenes (Buser, 1979) and the results obtained for the formation of polychlorodibenzofurans are also given. The chemical relationship between the precursors and the dioxins is obvious.

Figure 5 shows the formation of polychlorinated dibenzofurans upon pyrolysis of polychlorodiphenylethers (Lindahl et al., 1980). Several chlorodibenzofurans can result from the pyrolysis of one chlorobiphenyl. By realizing the number of PCB isomers in technical products and also in our present environment, there is also the appreciation that the combustion of products containing PCB could provide the source for all types of chlorodibenzofurans. One could argue that these more or less artificial experiments do not result in any environ-

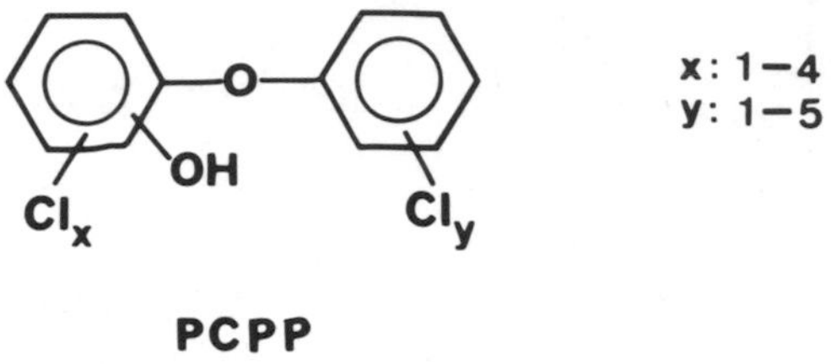
OH
Cl$_x$
Cl$_y$
x: 1–4
y: 1–5
PCPP

Fig. 3. Phenolic impurities in technical PCP.

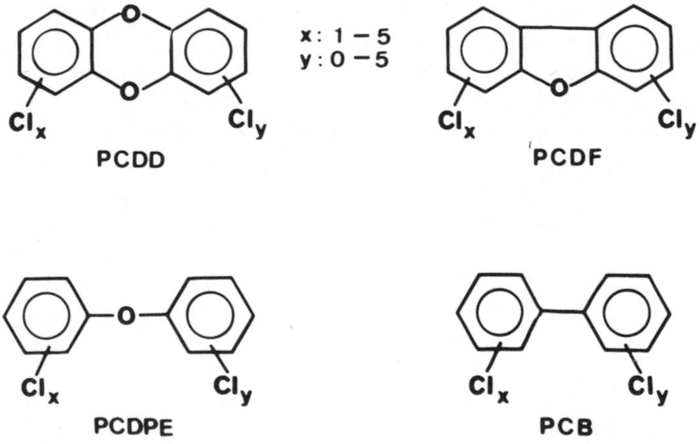
x: 1 – 5
y: 0 – 5
PCDD
PCDF
PCDPE
PCB

Fig. 4. Neutral impurities in technical PCP.

TABLE I. Formation of PCDDs from the Pyrolysis of Chlorobenzenes[a]

	PCDDs formed (ng/sample)				
Compounds	*Tetra-*	*Penta-*	*Hexa-*	*Hepta-*	*Octa-*
Trichlorobenzenes	30	20	<5	<5	<5
Tetrachlorobenzenes	<2	5	140	160	30

[a] *200 μg.*

mentally predictive data. However, when commercial chlorophenates are burned with respective impregnated wood, these chemicals also are formed.

Table II shows the tetra- to octachlorodibenzodioxins found upon the combustion of impregnated wood wool or birch leaves, respectively. As can be seen, the chlorodibenzodioxins are formed in microgram amounts upon the combustion of chlorophenate (Rappe and Marklund, 1978).

The same chemicals have also been identified in flue ash from waste incineration both under simulated conditions as well as from technical waste incineration. Based on this information it could be concluded that chlorinated industrial chemicals via their impurities or thermal conversion products are an important source of hazardous chemicals in the environment. Consequently, these derivatives should be included in the evaluation of the hazard potential of the different technical organochlorinated chemicals. It can be expected that there are also natural sources of these chemicals, although this has not yet been unequivocally demonstrated.

During the incineration of polyethylene under simulated conditions, in the presence of sodium chloride (Table III), no organo chlorine compound is present in the combusted material. However, a number of chlorobenzenes are formed. The concentrations given in Table III, based on the combustion of polyethylene, shows that if one ton of PE is incinerated about 15 gm of chlorinated benzenes result. It may well be that bush fires and the burning of wood have produced a number of chlorinated chemicals.

Thus, there is the realization that there are many sources of dioxins, furans, etc., and that although they are not universally present in the environment, they are still of some significance. Microbial degradation and sedimentation do not seem to play important roles in their elimination. Bioaccu-

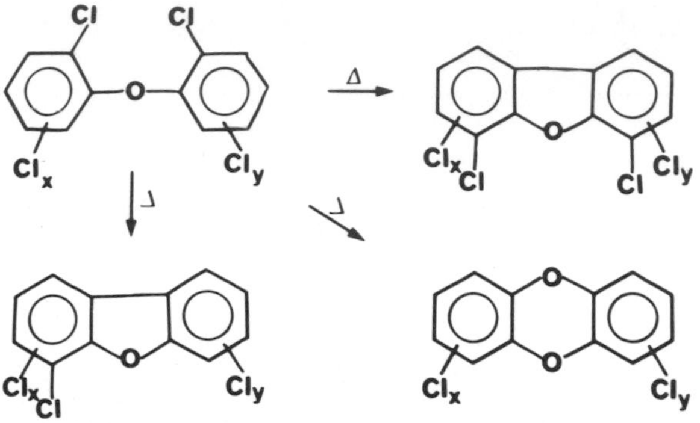

Fig. 5. Formation of polychlorodibenzofurans from polychlorodiphenylethers.

mulation is generally low; the accumulation is insignificant for daphnia and algae, which is well in agreement with the low water and lipid solubility and low partition coefficient in lipids. It seems that photochemical degradation could be the major route of environmental elimination (Fig. 6).

In the context of the Seveso accident and feasibility studies on decontaminating internal environments and, especially, soil, we investigated the photochemical degradation of 2,3,7,8-tetrachlorodibenzo-*p*-dioxin adsorbed on silica gel. With a wavelength of natural sunlight (dotted curve in Fig. 6) we found a rapid degradation of this most important dioxin, which indicates that at least the soil surface is decontaminated by sunlight. When using uv light, degradation was almost

TABLE II. Amounts of PCDDs Found in Combustion of Commercial Chlorophenates[a]

	Birch leaves		*Wood wool*	
Compound	*Charcoal*	*XAD-2*	*Charcoal*	*XAD-2*
Tetra-CDDs	35	26	96	210
Penta-CDDs	90	59	120	357
Hexa-CDDs	80	57	110	347
Hepta-CDDs	8	8	65	29
Octa-CDDs	0.3	0.2	1.2	1.2

[a] *μg PCDD/gm Servarex.*

quantitative within 4 days. Thus, uv irradiation could be used for indoor decontamination.

Although, there is much scientific interest in the continuous investigation of sources of dioxins, predioxins, etc., it can also be assumed they do not greatly change the environmental quality. On the other hand, despite all efforts it is difficult to assess these chemicals since they constitute a large number of chemical species which should be individually investigated for their ecological impact. Since they are not intentionally produced, and since there are many natural sources of these compounds, it is difficult to take any specific measures, except with respect to industrial accidents.

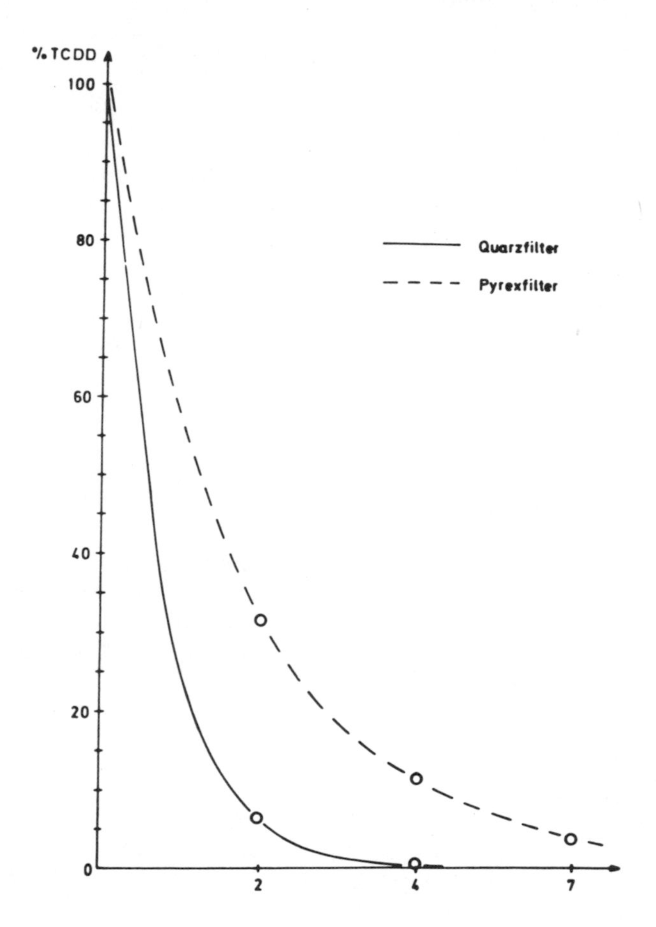

Fig. 6. Photochemical degradation of TCDD in 7 days.

TABLE III. Chlorobenzenes Formed by Incineration of PE in the Presence of Sodium Chloride

Chlorobenzene	*Concentration (μg/gm)*
Chlorobenzene	10
1,3-Dichlorobenzene	1.0
1,4-Dichlorobenzene	0.8
1,2-Dichlorobenzene	1.6
1,3,5-Trichlorobenzene	0.1
1,2,4-Trichlorobenzene	0.1
1,2,3-Trichlorobenzene	0.2
1,2,4,5-Tetrachlorobenzene	1.2
1,2,3,5-Tetrachlorobenzene	Detected
1,2,3,4-Tetrachlorobenze	0.2
Pentachlorobenzene	0.03
Hexachlorobenzene	0.005

From a scientific point of view it would undoubtedly be interesting to have at least some basic data on the environmental fate and on the potential effects of these species. From the point of view of the environment, efforts should be made to estimate the amounts and effects of dioxins, dibenzofurans, diphenylethers, and their respective polymers. All possible efforts should also be made to avoid the formation of these chemicals, even accidentally.

REFERENCES

Buser, H. R. (1979). Formation of polychlorinated dibenzofurans (PCDFs) and dibenzo-*p*-dioxins (PCDDs) from the pyrolysis of chlorobenzenes. *Chemosphere 3*, 415-424.

Buser, H. R., and Rappe, C. (1979). Formation of polychlorinated dibenzofurans (PCDFs) from the pyrolysis of individual PCB isomers. *Chemosphere 3*, 157-174.

Jensen, S. and Renberg, L. (1973). Various chlorinated dimers present in several technical chlorophenols used as fungicides. *Environ. Health Perspect.* 5, 37-40.

Pohland, A. E., Yang, G. C., and Brown, N. (1973). Analytical and confirmative techniques for dibenzo-*p*-dioxins based upon their cation radicals. *Environ. Health Perspect.* 5, 9-14.

Rappe, C. and Marklund, S. (1978). Formation of polychlorinated dibenzo-*p*-dioxins (PCDDs) and dibenzofurans (PCDFs) by burning or heating chlorophenates. *Chemosphere* 3, 269-281.

DISCUSSION

DR. BARNES: Is there any explanation as to why the methods used in 1973 to decontaminate the walls were not effective? Does the agent penetrate the coverings or what is the reason?

DR. KORTE: Painting over a contaminated area did not solve the problem because after a while these substances came through the paint. This is strange because dioxins are known to be immobile or become fixed to many materials and the mechanism of transport must be recognized in some way. There is no question that photooxidation via sunlight or ultraviolet light is the best method of decontamination.

There were no long-term effects found in many studies on dioxin, and there has been no official position taken by the German government with respect to any ecological effects. If the concentration of dioxins is not too high, painting and similar single decontamination measures might be successful, but not if the level is as high as that found in Zone A in Seveso.

DR. POCCHIARI: Accidents do occur in chemical plants that are not investigated in terms of toxicological effects. There are many reasons for this. What happens to the people is not known.

DR. CROSBY: Although tetrachlorodioxin is of great importance, a major chemical of concern is hexachlorodibenzodioxin, because this substance is present in a pesticide in much larger amounts, up to many parts per million, and as an impurity in most commercial pentachlorophenols. It is

ISBN 0-12-193160-9

much more slowly degraded by uv light than TCDD, and shows the same properties of moving in the environment. It eventually is adsorbed to sediments and its mammalian toxicity is almost as great as TCDD. Therefore, other kinds of dioxins should be investigated.

DR. COULSTON: As Director of the Bureau of Toxic Substances Management of the New York State Department of Health, Dr. Nancy Kim will describe the contamination of a large multistory office building by transformer oil containing PCB's.

CHAPTER 10

BUILDING RECLAMATION AFTER DIOXIN CONTAMINATION BY PCB FIRES

Nancy Kim

Bureau of Toxic Substances Management
of the New York Department of Health
Albany, New York

Of the several problems in New York State, dioxins are but one. It is not known how much dioxin is emitted from industrial stacks. PCB fires are very important. Transformers usually contain an oily mixture that includes PCB. If there is incomplete combustion during a fire, dioxins, chlorinated dioxins, and many other substances are formed. Although several of these fires have occurred, the one in Binghamton is of particular significance.

About 5:30 AM on February 5, 1981, there occurred a hot electrical fire in a switchbox in the basement of the State Office Building. A nearby transformer, which contained about 1100 gallons of liquid pyronox (a 55% mixture of PCB's and 35% chlorinated benzenes) was damaged by the fire. Approximately 180 gallons (655 liters) of the pyronox was lost. How much of it burned and how much leaked out before it was cleaned up is not known. As an immediate result of the fire, a large amount of soot was formed, which contained between 3 and 10 parts of PCB's as well as dioxin. The level of 2,3,7,8-TCDD in the soot was approximately 3 ppm from an output estimated to be about 200 ppm, thus indicating fairly high levels.

ISBN 0-12-193160-9

Most of the soot traveled from the basement area up through air shafts to other floors, to be deposited on walls, ceilings, pipes, appliances, desks, and so forth. There was great difficulty in proceeding with a clean up program, including the disposal of contaminated water. Within the building we could not flush the toilets for fear of discharging contaminants into the sewers. Therefore, we installed our own water treatment system. The contaminated water comes from the building and is held and then treated with activated charcoal.

Thousands of dry-wipe samples of the soot in the building were taken. The most highly contaminated areas were the spaces above the ceiling and in the walls. Levels of PCB found in these areas ranged from 600 to 16,000 $\mu g/m^2$ with an average of about 2,000. Levels found after cleaning were about 2 to 173 $\mu g/m^2$, with a geometric mean of 11 $\mu g/m^2$ compared to a precleaning mean of 162 $\mu g/m^2$.

New York State requires a permit for waste water disposal, one for ventilation of a contaminated building, and an environmental impact statement. Nevertheless, many of the problems that arose from the fire and subsequent ventilation would have been prevented by systematic planning.

As to the workers who have gone into the building, about 160 of them, who spent a great deal of time there, have been followed-up epidemiologically by the State Health Department. The first people that went into the building did not use proper protective equipment.

DISCUSSION

DR. BRAUN: There is little information on the fate of dioxins, especially on half-life, in man. Have you done any half-life or blood level measurements on the people who were exposed?

DR. KIM: Blood samples were taken and are being analyzed for PCB's and dioxins by the Health Department.

DR. AYRES: Was dry-dusting tried as a means of cleaning objects, especially since carbon is a good adsorber of dioxins?

DR. KIM: Dry cleaning by dusting was suggested but not tried, as a specific decontaminating method, although dust collected from the building was found to be rich in PCB's. Apparently, no one has tried or found a satisfactory dry cleaning material or method beyond use of a vacuum cleaner with a filter.

DR. CROSBY: I would like to ask about the levels of contaminants found on various surfaces such as desks, and whether or not there could be agreement reached on what was an acceptable level of dioxin(s) and/or dibenzofuran(s). Surely there must be some level of these various substances that would be considered safe for man and perhaps in many cases we have already reached these levels. If it is not possible to come up with some such agreement then one must start to demolish the buildings in this instance. In any case, it must be agreed upon that some level will be reached, as a result of the cleanup after these accidents,

ISBN 0-12-193160-9

that will then be considered safe and people will be allowed back into the building.

DR. POCCHIARI: In the case of the Seveso accident the level of contamination was reported to be 0.001 μg per square meter on the walls inside the houses. When this condition was reached the walls were painted and people returned to their homes and so far there have been no problems. Sometimes the total costs of cleaning these buildings may exceed the value of the buildings.

DR. KIM: There were no toxic effects observed in the approximately 160 people who were exposed to some degree in the Binghamton accident. There have been numerous accidents and fires involving the thousands of transformers in use and only isolated cases of apparent harm. These have not been documented, and until recently, there was no reason to suspect toxicity. Obviously, some important societal decisions are going to be made in the near future. On the one hand, there is a large economic problem; buildings cannot just be torn down, because there is no place to put the debris. On the other hand, the level of risk involved for people may or may not be as great as some authorities suspect.

DR. SILANO: Before we go on, I would like to give you some figures on what was considered as acceptable levels of dioxins in sub-Zones A-6 and A-7, because these questions have been put forward. The part of Zone A that was reclaimed lodged approximately 490 people, about 60% of the total population evacuated after the accident in Seveso. The number of houses and buildings was on the order of 200. For this work the authorities of the region of Lombardy established some maximal permissible TCDD levels to be the goal of decontamination. In the 7 mm of the top soil layer, a level of 5 $\mu g/m^2$ had to be reached. On exterior building surfaces the permissible level was 0.75 $\mu g/m^2$. On interior building surfaces the permissible level was 0.01 $\mu g/m^2$. The reclamation work consisted of washing the non-absorbant surfaces of buildings with water and a surface-active agent. Some floors were scraped. Painted walls were repainted with a synthetic varnish. After all this work was done we could not get any measurable amounts of TCDD. So that even the permissible levels were not indeed present. Of course, this assumption is within the limits of the detection in the monitoring system. The analyzed samples gave us a 99.5% level of confidence that the residue contamination, if any, would not involve more than 5% of the total surface of the buildings. If this was the case on the average, it was close to the maximum tolerated permissible level of TCDD I mentioned before. The population was very conservative. They would not agree with destroying all the houses and buildings, even if they were to be rebuilt in better shape. The point is that in situations such as this, people want to be very conservative and to go back to the situation the way it was before. We believe that in the case of the Seveso accident everything was done safely and this was the approach that was taken.

CHAPTER 11

MECHANISMS OF CARCINOGENESIS RELATED TO TCDD

Albert C. Kolbye, Jr.

Bureau of Foods
Food & Drug Administration
Washington, D. C.

The current dogma says that if there is a statistically significant increase observed in the incidence of cancer of test animals as compared to control animals, the substance is a "carcinogen." This is a quantitative observation which in the United States has become a qualitative philosophy. It is the author's opinion, that one cannot use a quantitative observation to prove a qualitative inference.

A better hypothesis might be that $I = \frac{V}{R}$. If I stands for incidence, V for effective potency, and R biological resistance, this hypothesis can be related to any disease that has a causative agent. The incidence of the disease reflects an interaction between a challenge to and the biological resistance of the organism involved. This applies to bacteria being used for mutagenesis testing, it applies to mammals being used for lifetime feeding assays, and it certainly, in the author's judgment, applies to humans.

Any factor that will function to increase the ratio on the right-hand side of this equation will appear as a relative in-

ISBN 0-12-193160-9

crease in the incidence. In terms of carcinogenesis, we know that some chemicals are genotoxic; they are electrophilic chemicals that appear to have a relative proclivity for alkylating or otherwise chemically reacting with DNA, and damaging it to the extent that replication of accurate genetic information is impaired or misinformation is introduced. We know that many chemicals are not active electrophilically until they have been activated. Semantically, they will be called precursors. If there is enzyme induction or a critical change in the metabolic pattern, the result may be increased activation of an electrophilic precursor, and a resultant increase in the bioconcentration of a genotoxic agent near the biological target. Any chemical effect that increases the activation of precursors may well appear as a carcinogenic effect if one has a biological system with certain initiators or potential initiators present.

There are other factors that can modify risk and thus the incidence of cancer, and appear as "carcinogens." These are disturbances of hormonal balance, nutritional aberrations, the presence or absence of viruses, and, of course, genetic factors that may be involved in a particular strain or species susceptibility.

Toxicants can also be considered. Our body has defense mechanisms against cancer induction. If they are impaired and there is present an independent causation system that potentially could become operative, then an increase in the susceptibility of that organism to the carcinogenic process will be observed. These impairments of defenses against cancer may involve disruption of membranes, enzymes, and substrates. Certain chemicals, when given in effective doses to selected organs, will cause a hyperplastic response and toxic hyperplasia, and, in many instances, will mimic some of the well-demonstrated changes associated with tumor promotion. The point should be made about toxic hyperplasia in contradistinction to classical tumor promotion is that toxic hyperplasia can increase the susceptibility of an organism to the induction of cancer *before* the addition of genotoxic carcinogen, whereas, a tumor promoter, a classical tumor promoter, is ineffective prior to administration of the initiator. A classical tumor promoter can only work *after* cells have been initiated, whereas toxic hyperplasia seems to increase susceptibility.

The equation already given can be written more correctly mathematically as functions. The incidence of cancer, I, equals the function of $VAPMT$/R. In the sorting out of some of these factors, V stands for electrophilic initiators, A for activators, P for potentiators, including the classical promoters, M for the modifiers, T for toxicants, and R for biological resistance.

In the normal mode, DNA will go through replication, with error-free enzyme repair activities. The result is normal replication of cells, i.e., tissues maintain their status. In carcinogenesis, a variety of things can occur. One of the first is a critical level of damage to DNA by genotoxic mechanisms. Genotoxic effects can be exerted by certain viruses and by radiation and is not limited to damage by chemicals. If there are also abnormalities of replication and repair in the presence of genotoxic damage to DNA, the probability of obtaining initiated cells will increase. An initiated cell is a cell that has now incorporated into its replicating genome a critical amount of DNA damage. This damage may be locked into the character of that cell forever, as long as that cell line is reproducing and surviving. An initiated cell, per se, does not mean a cancer.

For example, nobody knows yet how to accurately detect and measure initiated cells. We have every reason to believe that in liver, for example, there is a high population of initiated cells that will never, unless otherwise stressed, go onto cancer. It might be asked, what is the initiator? One of those initiators as a group might be the nitrosamines that are formed endogenously in mammals, if you accept the available biological evidence which, to the author, seems rather convincing.

Thus, if we have an initiated cell and then stress it with a continuous, consistent, selective pressure on that cell line (which is a way of describing what occurs in classical tumor promotion, or toxic hyperplasia), it seems to take about 15 cycles of cell reproduction (15 generations) for the cell to lose its normal biological controls and enter a mode that is described as a malignant cell. In other words, the cell becomes progressively recalcitrant in obeying any of the normal homeostatic biological controls in the body that are associated with regulating cell behavior, such as membrane surface phenomena, hormonal balance, etc.

A critical amount of DNA damage may be enhanced by other factors which increase the susceptibility of DNA to damage, and which will enhance the number of initiated cells. If we interfere with DNA repair or disrupt histones, or other factors which protect mammalian DNA, more initiated cells are obtained. If these replicating cells are repetitively exposed to toxic hyperplastic agents or promoters, the likelihood of cancer will increase.

The point is, cancer will not occur unless the promotional phase has occurred. Complete carcinogens, the highly electrophilic carcinogens, not only can initiate, but also are capable of self-promotion. Radiation itself, given at levels higher than those required to initiate cells, causes a form of toxic hyperplasia which is a promoting influence.

When tissues are damaged, putrescens, polyamines, and proteases are released. They can alter DNA and change its configuration by altering the histones and other nucleoproteins that act as supporting and protective structures to DNA.

Besides the histones, polyamines, and proteases, other alterations in phosphodiesterases and changes in membrane receptors occur. All of these are involved in the phenomenon of tumor promotion, which has been described as a chain of events known as the *cascade phenomenon*. Many of the biological parameters that are involved in promoter-induced cellular changes have also been demonstrated to occur in toxic hyperplasia.

In rodent liver, recent studies demonstrate that progressive changes in enzyme-deficient foci appear as a result of repeated administration of tumor promoters. An interesting marker change is obtained in glucose-6-phosphate ATPase and γ-glutamyltranspeptidase. There are also changes in iron stain. Also quite familiar is the Gary Williams' testing at the American Health Foundation. Some of these same changes appear to occur in the so-called "regeneration" nodules, which is an older term for one manifestation of toxic hyperplasia in the liver. Many of the chlorinated or otherwise halogenated compounds have a very low proclivity for electrophilic activity, but do have a capability of inducing similar effects in the liver. The changes are parallel to those of tumor promotion, and foci appear that progress to lesions, the end stage of which are hepatocarcinomas or hepatomas. Most halogenated compounds are indicted as "carcinogens" because of their effects on liver with respect to the induction of hepatomas and hepatocarcinomas. If it is borne in mind that liver probably always has a population of cells that are initiated independently by some other cause, then the application of consistent toxic pressure or consistent promoter pressure on liver will then induce liver tumors. The evidence that is available on dioxins must be considered in a broader context. Some generalities may be made.

In an earlier study, the two end points of concern were primarily liver and, to a secondary degree, lung. At the daily feeding level of one nanogram per kilo of body weight for a lifetime in rodents, there was a zero tumor effect. At 10 ng/kg of body weight by lifetime feeding, changes in liver that were alluded to above were observed. At 100 ng/kg of body weight per day, for a lifetime, frank liver cancer was seen as was the appearance of tumors in the lung. There have also been some comments about the biological half life of dioxins that have been discussed elsewhere. In general, it is fair to say that the biological half life of 2,3,7,8-TCDD in rat liver is approximately 25 days.

If the effect of daily dosing and cumulative effects until a steady state is attained is considered, then some rather enormous dosing of the liver is being dealt with.

As noted by Dr. Neal in another chapter in this volume, we do not know exactly (except for some of the binding characteristics) what is really happening in liver with respect to dioxin toxicity. Some of the information presented here provides some insight into what more to look for. If we are going to use rodents to indict chemicals or any other relevant biological evidence, then we also should use that evidence to talk about safety. The question was raised how safe is safe with respect to dioxin toxicity. If we believe the rodent feeding studies and transpose directly to humans, that means that the average adult human could tolerate feeding of 70 ng of 2,3,7,8-TCDD per day without any tumor effect.

This point is made because the social policy in the United States with respect to carcinogenesis has been to evaluate risk by using extrapolative techniques to linear risk extrapolation, or multistage models, etc. Thus, there is a very ultraconservative point of view. The author is willing to use ultraconservative models if they are indicated.

Linear extrapolation may have a place for highly electrophilic, highly genotoxic compounds. Conversely, linear extrapolation has little meaning for secondary carcinogens. In other words, dioxins are not primary carcinogens -- they are secondary. They are not electrophilic. We have some biological evidence about the dose range between no effect, intermediate effect, and frank, malignant effect, referred to in the Dow studies.

We need to think a little more about carcinogenesis in perspective and to investigate parameters known to be related to tumor promotion or toxic hyperplasia in relation to dioxin dose. We can then attempt to monitor humans to see if any of these changes have occurred from very intense accidental exposures to dioxins. The only human toxicity that the author is aware of with respect to dioxins is chloracne, having followed polychlorinated biphenyls, polybrominated biphenyls, and similar compounds for over 10 years. Occasionally, there is a suggestion of low level liver enzyme changes. However, for this class of compounds, humans have really never been stressed to the point where they have incurred any significant risk from relatively low levels of exposure.

DISCUSSION

DR. COULSTON: An important point to make is that for years we have been led to believe, by so-called experts on carcinogenesis, that any bump on the liver and any change in cell structure shall be called cancer. Apparently, there are many substances that really are only promoters and not direct carcinogens. One example from the laboratory is going to become classic. This is a compound known as 2-nitropropane, which is relatively innocuous for man, but for rats is very toxic and produces in adequate inhaled doses true hepatic carcinomas. However, it is not toxic to mice, dogs, cats, or monkeys. In the case of the rat, if the dose is lowered to the point where there is no hepatic cell damage that can be detected by microscope, alkaline phosphatase, or transaminase studies, cancer in these low-dose animals is never obtained in a lifetime study. The absolute no-effect level is 25 ppm. At 100 ppm one begins to obtain liver cell damage. At daily doses of 200 ppm by inhalation, the liver cells are destroyed within 4 months and cancer occurs. This suggests that we are dealing with another factor that may be considered a "promoter." When the liver begins regenerating new cells, as can be seen on the tissue section slides, these cells lead to cancer; this may, indeed, be considered a promotion effect.

DR. KOLBYE: These findings are consistent with observations made in a variety of other experiments with several compounds that are similar in character. The Dow Chemical Company has contributed to experimental studies in this area. The information gained from these investigations

ISBN 0-12-193160-9

will further explain these obscure mechanisms as they become more widely publicized. I hope that this knowledge will underpin a new approach to risk assessment by regulatory people in several agencies. One should distinguish between the effects of occasional and frequent exposure. For example, in the case of human exposure to dioxins that may be found in fish consumed only occasionally, considerable recovery may indeed take place between exposures. This kind of potential toxicity is different from the animal experiment where the rat is "pounded" day after day with some high level of the test substance. In the case of rats, all recovery is eliminated if a critical dose level is fed. This relentless pressure forces or pushes them right into cancer. The point is that the exposure patterns in these two cases are different and the risks are automatically different. The end result is that, if the exposure is less, the risk is less.

PROFESSOR TUCHMANN-DUPLESSIS: What is the relevance of the research you obtain in the rodent? What is the influence of the diet in the development of cancer in the rodent? Finally, what is the possibility of translating rodent data to man?

DR. *KOLBYE:* These problems are well known. The rodent is a most useful test system. Diet and the solvents used, all influence the results. The bottom line is that these animal test systems are the best that we can do at this time.

DR. *BRAUN:* Actually, the rat is a conservative model. The complications arise when we try to understand the significance of tumors that seem to be produced in animals by less potent carcinogens, the promoters. The mechanisms of action of these weak-acting materials, the promoters, must be clarified by the newer knowledge that is now being developed. At present, one can only speculate.

DR. *SILANO:* Do you feel that at this moment we can safely establish a no-adverse effect dose level for TCDD as far as long-term toxicity effects are concerned? If the answer is yes, what is the exact figure?

DR. *KOLBYE:* According to my interpretation and those of my associates at FDA, a comfortable no-adverse effect level is of the equivalent of 1 ng/kilo of body weight, which is 0.001 μg. That is the lowest level. There may be a need for further evaluation and interpretation of reproduction effects in order to become more comfortable with these figures.

DR. LENG: I would like to answer the question about the toxicity of TCDD in combination with other things. Workers at Dow have fed 2,4,5-T containing less than 4 ppb of total dioxins to rats for 2 years without carcinogenic effect at doses of 3,10, and 30 mg/kg per day. In another study in Germany, workers fed a technical grade of 2,4,5-T containing 0.05 ppm of TCDD and even at 30 mg/kg per day there were no carcinogenic effects after feeding for 2.5 years. This later study was a two-generation study. The rats were fed this diet prior to mating, through pregnancy, weaning, and, subsequently, for 2.5 years.

PROFESSOR TUCHMANN-DUPLESSIS: Professor Bruzzi will provide the details of a study of possible birth defects in the TCDD polluted area. Before the first drug law, from 30 to 40% of infant mortality was related to disease. For every 100 children born, 2-3 were reported to have congenital malformations. Actually, the number of malformations is much greater, but these arise after birth. There are also some malformations that arise only in later generations. There is general agreement that from recognized pregnancies in women, 60-75% end in spontaneous abortion. In some studies the incidence of malformation in therapeutic abortions is very high, perhaps 100 times higher than in the newborn. This is to say, if this mechanism of continuous screening were not taking place, the malformation rate in living humans would be not 5 or 6%, but at least 20%. Thus, in technical terms, understanding of the mechanisms by which the continuously deformed concepsus is eliminated, can be done by prenatal examinations and diagnosis with modern methods like continuous prenatal tests. This could be followed by therapeutic abortions.

CHAPTER 12

TERATOLOGICAL CONSIDERATIONS ON DIOXINS

H. Tuchmann-Duplessis

Membre de L'Académie Nationale
de Médécine
Paris, France

INTRODUCTION

Many data on dioxins have already been presented in this volume. Additional information is provided by Dr. Bruzzi in this volume. This chapter examines the experimental approach of teratogenicity and the safety evaluation of chemicals in the light of animal data obtained with 2,4,5-T and TCDD.

Until recently, the medical profession was mainly concerned with the teratogenic danger of drugs, which was exemplified by the "harmless" sleeping pill, thalidomide. This drug was responsible for several thousands of severely deformed babies. More recently, it was also established that environmental chemicals, such as, methyl mercuric sulfate, methyl mercuric chloride, and polychlorinated biphenyls were also capable of impairing prenatal development and causing congenital malformations in humans.

Since more women are professionally exposed to chemicals, to industrial wastes, and to a large variety of pesticides,

ISBN 0-12-193160-9

the potential hazard of these agents for their progeny must be considered. Among the 106 million American women, 43% are involved in some sort of professional activity. If 63% are at child-bearing age with a fertility rate of approximately 74 births per 1000 women, a yearly estimate of 2 million pregnancies can be made. Therefore, several thousand pregnant women could possibly be exposed to a dangerous environment. Since it is well established that a genetically normal ovum can develop into an abnormal embryo when exposed to noxious agents, the removal of such agents from the environment of women of child-bearing age is of utmost concern for teratologists.

THE MEDICOSOCIAL ASPECT

As a result of the advances in medicine, the main cause of infant mortality, namely, infectious disease, has been practically eliminated. Vital records in western countries show that over the past 40 years, the stillbirth rate has been reduced by 68% and infant mortality by 71%. The mortality of children between the ages of 1 to 9 years has been reduced by 87%.

In contrast to the relative safe environment of the newborn, the embryo faces more dangers than ever before. The use of very active drugs, disinfectants, and pesticides represent a potential risk for the growing conceptus.

Approximately 3% of newborns are obviously malformed. When these children are examined at ages 3 or 4, more subtle anomalies are detected. Therefore, the malformation rate raises to approximately 6 or 7%. Congenital malformations represent a major medicosocial problem because they are also the main cause of perinatal mortality and postnatal morbidity. More infants are handicapped for the rest of their life through developmental impairments than by any other cause.

The National Foundation has estimated that 7% of Americans suffer from some type of congenital defect and that seven hundred thousand infants are born defective each year.

PRENATAL DEVELOPMENT

To provide an overview of the factors involved in the causation of congenital malformations, it might be appropriate to summarize the pecularities of prenatal development.

During embryogenesis, there is a constant interaction between genetic and environmental factors. The genetic information coded in the DNA contains a program for the entire phenotype of the fetus, while the environment provides the nutrients necessary for the growth and differentiation. Hence, the expression of the inherited genes depends on exogenous factors.

A major characteristic of intrauterine development is the existence of critical periods. Although very short, the cellular activity is so dynamic during these periods that minor exogenous modifications can lead to major congenital defects. From the moment of exposure to exogenous factors, starting from gametogenesis, the effects on the conceptus are variable. They range from infertility to embryotoxicity, gross congenital malformations, and a large variety of more subtle morphological, biochemical, and functional abnormalities.

THE MORPHOGENETIC STAGES

The most critical phase starts with gastrulation, which in humans is at the end of the second week. This is a period of rapid morphogenesis regulated by precise and complex induction mechanisms which can be easily disturbed. Major malformations can be caused during this period.

Each organ passes through a period of maximum sensitivity; the heart between the 20th and 40th day, the central nervous system between the 13th and 25th day, and the limbs between the 24th and the 36th day.

The fetal period begins from the third month. This is a period of maturation and perfecting for the organs created during the period of embryogenesis. The external genitalia, which differentiate during the third month, and the central nervous system, whose histogenesis continues throughout the entire pregnancy and continues even after birth are most susceptible to exogenous agents. The central nervous system behaves like an organ system in which the various cells have a different multiplication and differentiation schedule. Therefore, according to the stage of the insult and the duration of the exposure, the morphological and behavioral damage varies. Certain alterations occurring during the late intrauterine stages express themselves only postnatally.

FRAGILITY OF THE CONCEPTUS

Among the causes of the high susceptibility of the embryo, is the lack of development of the enzyme systems necessary for detoxification of chemicals. Most of the oxidative enzymes which play an important role in the metabolism of various steriods and other drugs only gradually reach their normal values after birth. The poor ability of the fetus to form glucuronides also enhances the susceptibility of the nervous system to various chemicals, including drugs. The immaturity of the fetal kidney constitutes an additional handicap to the fetus, since it cannot detoxify and excrete the products or metabolites as rapidly as the adult.

In the human fetus, however, the enzymatic activity is more developed than in experimental animals. Some species differences, as the lower susceptibility of the human embryo, might possibly be related to such enzymatic peculiarities.

PREVENTION

The moral and financial burden of congenital malformations is enormous. Prevention should, therefore, be of utmost importance for scientists. There are two approaches to this problem: clinical and experimental. The clinical approach takes into consideration the spontaneous elimination of abnormal conceptuses. It is estimated that approximately 15-20% of recognized pregnancies end in spontaneous abortion. Morphological examination of such abortuses reveal that in more than 60% of these cases, obvious abnormalities are found.

Nishimura et al. (1966), who examined 3000 human embryos from 3 to 10 weeks of age, found that the frequency of certain malformation like spina bifida was ten times higher in embryos than in neonates. A somewhat similar situation was also found for chromosomal aberrations. Would such a mechanism of elimination not exist, the malformation rate in humans would not be 3% but probably three times higher.

The spontaneous defense mechanism can be improved by early prenatal diagnosis of malformations using amnioscopy, amniocentesis, ultrasound scanning, direct visualization by fetoscopy, and other modern techniques, followed by induced abortion, if the birth of a defective child can be foreseen.

The experimental approach, which is the most dynamic, aims to identify exogenous agents capable of impairing intrauterine development with the view of removing them from the environment of pregnant women. Teratologists have developed screening methods for the safety evaluation of drugs, food additives, and pesticides, which have contributed to decreased usage of potentially dangerous compounds. Such experimental investigations have also been performed with two compounds, e.g., 2,4,5-T and TCDD, which have been incriminated in the etiology of congenital malformations.

ANIMAL DATA ON DIOXINS

High doses of 2,4,5-T containing 0.1 ppm of TCDD are embryotoxic and cause cleft palate in mice. Using doses of

3 μg/kg, administered between the 6th and 15th day, Moore (1978) noted a very high percentage of cleft palates and hydronephrosis in mice. Neubert and Dillman et al. (1972) have made similar observations with doses of 1 to 10 μg/kg. Furthermore, they found a high embryotoxicity of dioxin, which is evident at doses which do not appear to affect the health of the mother. In addition to the embryotoxic and teratogenic effects, dioxin also induces adipose infiltration of the liver, intestinal hemorrhages, edema, involution of the lymphatic tissue and thymus gland, and slower bone formation in the fetus.

The action of the lymphatic system is marked by lymphocyte depletion. Vos and Moore (1974) found a suppression of cellular immunity in the offspring of both rats and mice when treatment was carried out during the second half of the pregnancy or during the postnatal period. Experiments in rats and in mice revealed that the highest no-effect dose level in rats was 0.03 μg/kg/day. In monkeys, the no-effect dose level seemed to be 10 times higher. In fact, there are important differences in susceptibility to TCDD among various species. For example, the oral LD_{50} was found to vary from 1 to 1000 between the guinea pig and the hamster (Table I). The mechanism of the toxic action of TCDD is still not established. The acute toxicity for human is unknown.

A three-generation study in rats revealed that the neonatal survival through the 21-day-suckling period was decreased in a dose-related manner in the group that received between 10 and 30 μg/kg/day of 2,4,5-T in their diet.

TABLE I. TCDD Toxicity

Animal species	*Oral LD_{50} (μg/kg)*
Guinea pig	0.6-2
Male rat	22
Female rat	45
Monkey	70
Rabbit	115
Dog	30-300
Hamster	1000

In the group that received 3 μg/kg/day, no adverse effects were seen. The dosage of TCDD contained in the 2,4,5-T consumed by the 3 μg/kg/day group was 0.00000 μg/kg/day.

In an experiment by Kociba et al. (1976/1979) (see Table II), doses of 0.001-0.1 μg/kg/day TCDD were administered in the diet. At dosages of 0.01 and 0.1 there was decreased fertility and the survival of the young. In the 0.001 μg/kg/day group, no adverse effects were observed during the period of fertility, in the number of pups per litter, or in postnatal body weight gain. Survival was not modified.

Experiments carried out with primates, admittedly with relatively low doses, suggest that TCDD might be less harmful to the rhesus monkey than to rodents. Dougherty et al. (1975) administered the herbicide 2,4,5-T containing 0.05 ppm tetrachlorodibenzo-*p*-dioxin to rhesus monkeys between the 22nd and 38th day of pregnancy. Doses of 0.05, 1, and 10 mg/kg were not toxic for the mothers and no malformed offspring were produced for 1 year. These results suggest that the toxicity of 2,4,5-T may be a species-dependent phenomenon. McNulty (1977; McNulty et al., 1981) investigated the toxicity of TCDD and TCDF in rhesus monkeys. Dietary levels of 50 and 5 μg/kg caused a high rate of mortality.

The most obvious signs of toxicity are atrophy, squamous metaplasia of sebaceous glands, mucous metaplasia, and hyperplasia of the gastric mucosa. In addition, after 2 months of treatment, the body weight is decreased and the thymus and the bone marrow show obvious hypoplasia.

The reactions to TCDF are similar to those following TCDD ingestion. However, an important difference exists. With the former, rapid recovery follow interrupted exposure. This is not the case with TCDD-treated animals. The determination of an LD_{50} in rhesus monkeys is difficult for either TCDF or TCDD. However, it is clear that a few micrograms per kilogram, even when distributed for a few weeks or months, are lethal for most rhesus monkeys.

The influence of these chemicals on reproduction was also investigated. When rhesus monkeys were treated from day 20 to 40 of pregnancy with doses of 0.2 (Group I), 1 (Group II), or 5.0 μg/kg/day (group III), the abortion rate of the first group was 1:4; in group II, 3:4 monkeys aborted. In the last group, the two treated animals aborted. No obvious malformations were found among the recovered fetuses. In a tentative extrapolation to human exposure, McNulty considered that the maximum permissible level of TCDD was 2 ppt. This is a level which is below the sensitivity of the chemical assay. This estimate appears to be unrealistic and has been questioned since much higher doses did not impair reproduction in humans.

TABLE II. TCDD Effects on Fertility in Rats[a]

Dose (ug/kg/day) 90 days	F_0	F_1	F_2	F_3
0.001	No deleterious effects on fertility	No deleterious effects on fertility No effect on litter size at birth or on neonatal growth	No deleterious effects on fertility No effect on litter size at birth or on neonatal growth	No deleterious effects on fertility No effect on litter size at birth or on neonatal growth
0.01	No decrease of fertility	Fertility significantly decreased	Fertility significantly decreased Systemic toxicity clearly evident	Systemic toxicity Smaller litter size Decreased survival and growth of neonates
0.1	Fertility decreased to an extent which precluded continuation of the study No survival rate			

[a] *From Kociba et al., 1976.*

Lamb, Moore, and Marks (1981) investigated the action of mixtures of (simulated Agent Orange) 2,4-dichlorophenoxyacetic acid, 2,4,5-T trichlorophenoxyacetic acid, and 2,3,7,8-tetrachlorodibenzo-*p*-dioxin TCDD in male mice. These chlorinated phenoacetic compounds are widely used as herbicides in forestry and agriculture. It has been estimated that 107 million pounds of herbicide Orange, which was used as a defoliant, was sprayed in Vietnam during the years 1967-1979. In this experiment, two hundred mice were divided in to eight groups. One half (four groups of 25 mice) was used for toxicity evaluation, while the other half was utilized for fertility and reproductive studies. The testicular function was assessed by sperm counts, evaluation of motility, and morphological anomalies. No significant differences in sperm were found during or after the dosing period. After 8 weeks of dosing, treated males were mated to virgin untreated females. The mating frequency, average fertility rate, implantation and resorption sites, and percentage of malformations were measured in relation to the treatment. The average number of implants per litter was not decreased. There was no significant decrease in fertility. These results clearly show that in the exposed groups spermatogenesis was not impaired. Survival of the offspring and neonatal development were also unaffected by exposure of the male to these chemicals. It can be concluded from this study that a continuous administration of these compounds during the entire period of spermatogenesis does not increase productive abnormalities in the 2,4-D- and 2,4,5-T, or TCDD-exposed groups. The level of TCDD used in this study was within the range which causes cleft palate and kidney anomalies in mice. In addition, the exposure to these chemicals did not influence the fetal or neonatal development or the viability of the offspring sired by the treated mice.

In summary, animal data show that 2,4,5-T and its main contaminant TCDD, although embryotoxic, causes congenital malformations only in mice but not in other species. There is no evidence that exposure of male mice even to relatively high doses of chemicals impairs reproduction. The results observed in monkeys suggest that the normal use of 2,4,5-T does not constitute a potential teratogenic danger for humans.

Although most of the insecticides, fungicides, and herbicides are toxic in animals, intoxication has only occurred accidently in overexposed humans. There is no evidence that organo chlorine compounds when used under normal conditions caused intoxication in humans.

PREDICTIBILITY OF EXPERIMENTAL DATA

The main difficulty in extrapolating animal data to humans lies in the fact that various species have a different susceptibilities to noxious agents. This is particularly important in the case of TCDD since even the LD_{50} varies from 1 to 1000 according to the species used. Furthermore, it has been shown that the embryo of highly selected strains has a greater susceptibility than the human embryo. Besides the differences in susceptibility to TCDD, the concentrations used in experimental animals were much higher than those to which humans were eventually exposed. These two facts suggest that the normal use of 2,4,5-T should not constitute a potential teratogenic danger for the progeny.

CONCLUSIONS

The potential risk of environmental factors during intrauterine life is of particular concern because of the irreversible nature of such insults. A large variety of teratogenic agents have been studied in animals through well-designed experimental screening. To produce congenital malformation, the teratogenic agent not only has to be given at an appropriate dosage and at a very precise stage of the morphogenesis but, in addition, the embryo must have a suitable genetic susceptibility to be capable to react. All of these conditions can be readily achieved when experiments are performed on a large number of animals of a genetically defined stock. A combination of such complex circumstances is very rare in humans. Moreover, the human embryo seems to be far more resistant than animals used in these teratological experiments. Furthermore, in humans exposure does not reach the high concentrations which are used in animal experiments which are necessary to produce embryotoxic or teratogenic effects.

These facts probably explain the discrepancy between experimental results which have led to the discovery of a large number of teratogenic agents (more than 400). Clinical observations have only established the lack of toxicity to the human embryo for less than twenty external agents.

As more knowledge is gained by experimental investigations and human epidemiological studies, the evaluation of the teratogenic risk of environmental agents will become a possibility.

REFERENCES

Courtney, K. D., Gaylor, D. W., Hogan, M. D., and Falk, H. L. (1970). Teratogenic evaluation of 2,4,5-T. *Science 168*, 864-866.

Courtney, K. D. and Moore, J. A. (1971). Teratology studies with 2,4,5-trichlorophenoxyacetic acid and 2,3,7,8-tetrachlorodibenzo-*p*-dioxin, *Toxicol. Appl. Pharmacol. 20,* 396-403.

Dougherty, W. J., Herbst, M., and Coulston, F. (1975). The non-teratogenicity of 2,4,5-trichlorophenoxyacetic acid in the rhesus monkey *(Macaca mulatta). Bull. Environ. Contamin. Toxicol. 13,* 477-482.

Kociba, R. J., Keeler, P. A. Park, C. N. and Gehring, P. J. (1976). 2,3,7,8-Tetrachlorodibenzo-*p*-dioxin (TCDD): results of a 13-week oral toxicity study in rats. *Toxicol. Appl. Pharmacol. 35,* 553-574.

Lamb, J. C., IV., Moore, A., Marks, Thomas A. (1981). Evaluation of 2,4-dichlorophenoxyacetic acid (2,4-D), 2,4,5-trichlorophenoxyacetic acid (2,4,5-T), and 2,3,7,8-tetrachlorodibenzo-*p*-dioxin (TCDD) toxicity in C57BL/6 mice: reproduction and fertility in treated male mice and evaluation of congenital malformations in their offspring. In press.

McConnell, E. E., Moore, J. A., Haseman, J. K., and Harris, M. W. (1978). The comparative toxicity of chlorinated dibenzo-*p*-dioxins in mice and guinea pigs. *Toxicol. Appl. Pharmacol. 44,* 335-356.

McNulty, W. P. (1977). Toxicity of TCDD for rhesus monkeys: brief report. *Bull. Environ. Cont. Toxicol.*

McNulty, W. P. et al. (1981). Chronic toxicity of TCDD for rhesus macques. *Food Cosmet. Toxicol. 19*(1), 57-65.

Moore, J. A. (1978). Toxicity of 2,3,7,8-tetrachlorodibenzo-*p*-dioxin. *In* "Chlorinated Phenoxy Acids and Their Dioxins-Mode of Action, Health Risks and Environmental Effects," (C. Ramel, ed.), pp. 133-144. NFR, Stockholm.

Murray, F. J., Smith, F. A., Nitschke, K. D., Humiston, C. G., Kociba, R. J., and Schwetz, B. A. (1979). Three-generation reproduction study of rats given 2,3,7,8-tetrachlorodibenzo-*p*-dioxin (TCDD) in the diet. *Toxicol. Appl. Pharmacol. 50,* 241-252.

Neubert, D. and Dillmann, I. (1972). Embryotoxic effects in mice treated with 2,4,5-trichlorophenoxyacetic acid and 2,3,7,8-tetrachlorodibenzo-*p*-dioxin. *Arch. Pharmacol. 272,* 243-264.

Nishimura, H., Takano, T., Tanimura, M., Yasuda, T., Urchido (1966). High incidence of several malformations in the early embryos as compared with infants. *Bull. Neonate. 10,* 93-107.

Tuchmann-Duplessis (1975). "Drug Effects on the Fetus." Adis Press, New York, London.

Vos, J. G. and Moore, J. A. (1974). Suppression of cellular immunity in rats and mice by maternal treatment with 2,3,7,8-tetrachlorodibenzo-*p*-dioxin. *Intern. Arch. Allergy 47,* 777-794.

DISCUSSION

DR. COULSTON: Does organ specificity in terms of birth defects relate to the time of pregnancy or exposure to the toxic substance?

PROFESSOR TUCHMANN-DUPLESSIS: If you had a medium-grade teratogenic substance, it will be closely related to the time of exposure. If you have a very powerful teratogen, like actinomycin D, you get a wide variety of malformations.

DR. KOLBYE: I would like to point out that when one uses an expression "weakly mutagenic," it is a very understandable term, but by classical toxicity you can get what appears to be a low level of mutagenicity. However, it really is not mutagenicity but straight toxicity. Mutagenicity is measured by incidence phenomena in all test systems and the same factors mentioned with respect to carcinogenicity also operate in mutagenesis. This is often a cause of confusion.

ISBN 0-12-193160-9

OPENING REMARKS

DR. COULSTON: The man who will be your chairman is really a pioneer with wide experience in the field of toxicology. His skill and background in this area and his broad experience as an administrator make him well qualified to guide these discussions. I am pleased to have him here with us and I ask Dr. Frawley to chair this session.

DR. FRAWLEY: I certainly want to thank Drs. Coulston and Pocchiari for inviting me to chair this session. This is indeed a unique meeting and is typical of the many meetings in the area of toxicology organized by Dr. Coulston that prove to be most productive and fruitful in terms of effective content. I have been interested in the history of the dioxins, especially when so little was known about these potentially toxic substances that now seem to be less hazardous.

In the early 1960's the manufacture of 2,4,5-T, the finished products contained 3-5 ppm and some as high as 28 ppm of dioxins. The manufacturers at that time all had occupational health programs for their workers. Perhaps these programs were not as sophisticated as those of today, but certainly the physical medical examinations and related health observations were valid. There was no evidence of toxicity or even a single case of chloracne. This is especially significant in view of the fact that the people at that time were packaging 2,4,5-T containing several ppm of dioxin, in reasonably dusty operations, without protective clothing. In addition, there were many spray operators who were handling 2,4,5-T with a few ppm of dioxin without any special protective cloth-

ISBN 0-12-193160-9

ing. Again, their health record was good. Perhaps the concern about 3 ppm in the dust was greater than the facts warranted.

DR. KIM: One issue should be clarified. The contamination in the New York Office Building was measured only in terms of 2,3,7,8-TCDD and 2,3,7,8-TCDF and the amounts of these in the soot were as high as 200 ppm. Other isomers were also identified, so there was a conglomeration of various isomers, although attention was directed to the presupposed most toxic isomers.

DR. BRUZZI: The work to be discussed in the next chapter was the result of several teams of investigators who participated in the epidemiologic study following the Seveso accident.

CHAPTER 13

HEALTH IMPACT OF THE ACCIDENTAL RELEASE OF TCDD AT SEVESO

Paolo Bruzzi

Istituto Scientifico per lo
Studio e la Cura dei Tumori
Genova, Italy

I. INTRODUCTION

This chapter presents what is known on the health status of the population of the Seveso area in relation to the Icmesa accident, and to the potential TCDD exposure. This discussion is based on the same evidence available to all Commissions or Scientists working in Seveso, and reflects, in general, what is presently agreed upon by every researcher in Seveso.

The occurrence of the Icmesa accident has already been widely discussed in several articles and books from both a technical and chemical viewpoint; this aspect, therefore, will not be discussed. One feature which must be emphasized is the almost unique feature of this disaster: this is one of the few instances in which a wide, densely populated area has been heavily polluted by an extremely toxic substance, capable of affecting several organs and systems both acutely and chronically. The involvement of a general population and the etherogenicity at the types of exposure, acute and chronic, heavy and light, in every possible combination, has made it ex-

ISBN 0-12-193160-9

tremely difficult to design and carry out an epidemiological strategy.

Two factors characterize the history of epidemiological studies in Seveso: the lack of proper studies in the early phase, for reasons which will be explained later, and problems in defining this exposure. Regarding the former point, it is useful to illustrate the characteristics of the area in which the Icmesa plant is located. It is a densely populated area with a mixed social and economical structure: family agriculture, homeyard animals breeding, small handicraft shops and industries, wood and furniture factories, and chemical plants of various types. Together this exists in one of the wealthiest areas of Italy, and from where much immigration from the South and from the East has taken place during this century. The suddenness of the Icmesa accident has literally upset both the social and economical structure of this area. During the initial period, when animals were dying and children were developing skin problems, it was expected that the whole populations was going to show major health short- and long-term effects. The attitude which then evolved was a kind of a "self-defense," a psychological refusal for everything connected with dioxin. This attitude has caused and is still causing many difficulties in executing clinical studies. It is in this atmosphere that the too often criticized epidemiological studies in Seveso area had to be designed and started. This atmosphere was also reflected in the actions undertaken during the period immediately following the incident. Following the evacuation of the most polluted A Zone, an intensive screening program was started, in which blood tests and clinical examinations were freely offered to anyone living in the polluted area, or to those that had been potentially exposed to TCDD. An intensive campaign of chemical analysis of ground samples from a wide area for presence of TCDD was also undertaken. No standardized protocols or control groups were identified. The population was expected to show many effects and it was expected that refined studies were unnecessary.

During the summer of 1979, the Istituto became involved in the Seveso problems. It was already clear that, fortunately, a major disaster of population-size had not taken place. On the other hand, the lack of well-designed epidemiological studies had made it impossible to rule out that minor health effects had affected large portions of the population of the area or that very serious effects had occurred in very limited subgroups.

The long-term effects were still in question. In order to design an appropriate epidemiological strategy, two main problems had to be resolved: (1) Census of the study population and (2) identification of exposed and control groups.

The first problem has been resolved by a demographic register of the entire monitored area to establish the births, deaths, and movements of 220,000 people. This record has enabled (1) the calculation of rates by place of residence, age, sex, from 1976; (2) the tracing of persons moving from one part of this area to another; (3) the association of newborn data with mother data; and (4) the identification of persons by place and date of birth, which is necessary for the Cancer Register.

The second problem has yet to be resolved. Environmental TCDD exposure among the general population is not detectable through any reliable and manageable biological indicator, with the exception of chloracne, which is specific but not sensitive. The assumption that only children and a few adults had actually been exposed to TCDD in Seveso area appears to be unrealistic. Other possible biological indicators like liver enzymes, triglycerides, or indicators of enzymatic induction are not specific and are of unknown sensitivity. They are also often very expensive.

The parameter that has been chosen to identify exposed and controlled groups was not very reliable, but was the only one available and at practically no cost. It was established that for population-based studies of rare phenomena, like birth defects, cancer, and specific causes of mortality, in which large study groups are needed, the population of the whole area had to be divided by place of residence in zones at different pollution levels. At present we are comparing, in every study, rates or averages in Zone A and B, with Zone R (see Chapter by Pocchiari) and with the outside area, which were the zones of decreasing levels of pollution identified on the basis of chemical testing of soil samples. Unfortunately, for reasons which are still unclear, there is not perfect agreement between this zoning and the behavior of other parameters, as chloracne and other dermatological indicators, as well as deaths. An obvious declining trend was observed. Chloracne cases were observed in the R Zone and even in the out zone. The same was true for cases of acute dermolesions and atrophodermia. Similar findings were observed for animal death, animals which died of typical syndrome, and rabbits with TCDD in their liver. Our subdivision, therefore, provided only a rough indication of populations which, on the average, lived in an environment at different levels of pollution. These are called "ecological studies" and their results must be accepted cautiously.

The present strategy, however, is aimed at predisposing methodologically correct studies for mid- and long-term evaluations of these issues which, due to their social and health relevance, deserve special attention and require precise answers. Therefore, top-priority projects were selected related to specific causes of mortality, adverse pregnancy outcomes (i.e., abortions, stillbirths, and birth defects), and cancer incidence. As necessary tools were the demographic Register, the utilization of hospital admission-discharge forms, and the definition of risk maps which take into account the distribution of chloracne cases and animal casualties. Among the follow-up studies, special efforts were devoted to the area of occupational medicine, in light of the availability and feasibility of several studies related to this matter. Clinical studies of the general population were not undertaken since the accident had occurred 3 years prior.

II. MORTALITY

Mortality studies on a general population exposed to a chemical are not very powerful tools for detecting health effects, but are necessary for two reasons: (1) to establish a baseline, in which the general mortality pattern of that population is defined in relation to possible relevant deviations, and (2) as a source of data for organ-specific incidence or mortality studies (cancer registry, liver diseases, etc.).

Our data on mortality covers from 1975 to 1980, and the mortality pattern of Seveso area reflects that of industrial countries, with cardiovascular diseases and cancer as leading causes of death. Mortality from liver diseases is higher than in the rest of the Lombardy Region, but this phenomenon was found in 1975. No remarkable differences between zones at different pollution levels were observed, or was there acute deaths attributable to TCDD exposure in Zone A residents. There was one case of liver cancer death in 1977 but it was not attributable to TCDD exposure due to the Icmesa accident, because of a too-short latency period.

III. ABORTIVITY BIRTH DEFECTS

Among the most striking effects of TCDD on animals of particular importance is its teratoxicity and teratogenicity.

Human data are scanty. Data from Vietnam and the indirect evidence from the Swedish studies, which are concerned with hexachlorophene, are not too reliable. This is one of the few instances in which, with no doubt, exposure to TCDD has occurred during pregnancy.

It needs to be pointed out that, although this has been the area to which intense efforts have been devoted, there are several questions to which no answers are as yet available for the following reasons: (1) Lack of reliable indicators of exposure; (2) problems of sample size; (3) delay in the implementation of the studies; and (4) the results of the studies on abortivity and birth defects which started on separate protocols, have not yet been collated.

Nevertheless, these studies do provide important indications. The results of the studies on abortivity from 1976 to 1979 must be presented separately because the quality of the data and the reliability of the results are completely different: the first was from July, 1976 to December, 1977; the second was from 1978 to 1979.

A statistical analysis of the data immediately following the accident is not yet possible, because the input in the Desio Hospital Computer is still incomplete. The internal reliability of the data and the calculation on the progeny from exposed individuals have not yet been completed. Considering all these limitations the data for pregnancy from October 1976 to September 1977 show higher rates of abortivity in Zones A and B compared with that observed in the Zone R and in the out area, which are quite similar. In the third trimester 1976, many exposed women had induced abortions. Many pregnancies started outside the area because of holidays and evacuation. The abortivity rates observed in Zones A and B during these four trimesters were the highest observed in any period and in any zone.

The rates for the second period which are much more reliable, show a declining trend in Zones A, B, and R in contrast with stable rates in unpolluted zones. This trend does not reach statistical significance, but it is quite suggestive. A clear seasonal variation is observed both in 1978 and 1979 in the polluted zones but not in the out area. The difference between abortivity rates during 1978 in Zones A, B, and R versus the out area reaches statistical significance (A+B+R = 12.75%; out area = 9.82%; Z = 1.98 $P < 0.05$). This difference disappears in 1979.

In conclusion, the data available provide evidence favorable to an increase in abortivity rates in the polluted zones, from September 1976 to 1978, possibly attributable to the accident. Values returned to normal after 1978.

The data concerning birth defects, on the contrary, show a much more confusing picture. Birth defects, which appear, as a whole, with a frequency varying from 25 to 150 every 1000 newborns, are relatively rare when considered separately, with frequencies varying from 0.1 to 10-20 per thousand newborns.

Therefore, if we stratify the 2700 births per year in the whole area by pollution zone, we obtain very small numbers of little, if any, statistical meaning. Furthermore, before the accident, birth defects in this area were largely under-reported. Thus, we cannot compare our data with those of previous years. Comparisons with data from other registries are affected by possible differences in the various operations of each registry and by the geographical variation in baseline birth defect rates.

Most of our rates fall within international ranges, which provides the first indication of the reliability of our register. The high rate observed for Down's Syndrome conceals the fact that the rate is close to that observed in most Latin countries. The excess observed for hypospadias will be discussed later.

The Seveso rates were compared to the rates observed in various groups of Italian hospitals reporting to the Multicentric Survey of Birth Defects. The rates for esophageal and rectal anal atresa (one case each) and for limb reduction deformities (three cases) are below the reference ranges, probably due to chance variation.

An excess in neural tube defects was also observed. This excess is compared with the data of the Lombardy Region taken from the same source and is mostly due to spina bifida. The highly significant excess observed again for hypospadias is only partially accounted for by differences in diagnostic criteria. In fact, it is statistically significant ($P < 0.05$) if we consider only penile and scrotal hypospadias.

The Seveso birth defect register was started in 1978. This could possibly explain the small number of cases observed among newborns of 1977. However, the extent of under-reporting does not affect all types of malformations. If we exclude hemangiomas, overall malformation rates show a clear declining trend from 1978 to 1980. There is a declining trend for cardiovascular and, possibly, for musculo-skeletal and genital anomalies.

Higher rates in polluted zones can be observed for some groups of malformations, such as those affecting heart and vessels, the genital system, the integument, and other identifiable syndromes.

Children with multiple malformations by year and by zone of occurrence have been calculated. Even though the small number of cases precludes a meaningful statistical testing, the correlation between potential TCDD exposure and relative risk is suggestive.

Of the observations presented with regard to those anomalies which have shown some remarkable feature in our Register, the possibility of an association with TCDD pollution is not supported by the overall evidence of polydactily and Down's Syndrome. An association is suggested by the available evidence concerning hemangiomas and, perhaps, neural tube defects, and it is more clearly indicated for hypospadias, because of a strong consistent excess over other registries and a weak correlation with potential TCDD exposure, and for both cardiovascular and multiple defects, because of a reasonably good correlation and a possible time trend. Further studies are needed to verify such hypotheses.

IV. CANCER INCIDENCE

Due to the latency period, cancer incidence studies were not very urgently needed, but it was important to set up registration procedures in order to obtain stable and reliable baseline rates.

Data for the residents in 11 municipalities of the CSZ 1, 2,3 of Brianza of Seveso, hospitalized for cancer in one of the hospitals of the Lombardy Region from 1975 to 1980 were collected. These data were derived from the hospital-discharge diagnosis that were routinely stored in a computer for all of the Lombardy Region.

The main problem which now exists is to verify the diagnosis of cancer and to connect this with specific individuals in order to avoid inclusion of multiple hospital admissions for same case in this data. The data was collected by checking the diagnosis of several selected types of cancer, cancer among children under 15 years of age, and other types of cancer that can be considered more interesting on the basis of the animal experiments, epidemiological data or, possibly, a shorter latency period. Although only few preliminary manually processed data are available most of the cancers were considered "rare." It is not known whether, and how, at this point, our data reflects prevalence and/or incidence rates. In general, the data reflect the rates observed in the Registry of Varese, a town close to the Lombardy Region.

V. CLINICAL STUDIES

The problems previously discussed also severely affect the validity of the results obtained with the clinical studies: (1) Poor population compliance at scheduled examinations and tests. (2) The lack of a control group and the inadequacy of standardization both in laboratory and clinical procedures.

The problem of compliance reached a satisfactory level in the initial period following the accident only for Icmesa workers and perhaps for Zone A inhabitants. Even in Zone A, over a 4-year period, the number of tests per person varied from 1 (10%) to 10 (7%). It was observed that subjects with at least one abnormal test were more frequent among chloracne children and the Zone A population. However, these two groups complied with the more frequent examinations.

Similar problems were present in clinical medicine studies, which covered only chloracne cases, A Zone inhabitants, and those persons with pathological findings attributable to TCDD. However, bearing in mind these limitations, the following evaluations were noted in the report on public health status by Professor Zanussi, of the Government Scientific Commission: "393 individuals, among those monitored, were found with clinical or laboratory abnormalities potentially associated with TCDD intoxication. One hundred forty-eight showed increased urinary porphyrins, 35 of whom with evidence of liver disease. Three hundred eighty-nine had increased cholesterol or triglycerides, 255 of them with evidence of liver cirrhosis, 90 of chronic hepatitis and 25 of platelets deficiency were found. The most commonly affected tests were GGT, and alkaline phosphatase, while transaminases showed an increase in the second half of 1976." The report is concluded as follows:

1. The incidence of clinical hepathomegaly, with all the reservations that can be made in this regard, is high especially in individuals coming from Zone A and Health District Number 2.
2. The same applies to some laboratory indices of impaired liver function.
3. Some other laboratory tests, such as those for serum cholesterol and triglycerides, were often altered.
4. Other changes (microhematuria, leukopenia, hypogammaglobinemia) were found in a small number of individuals.
5. The alterations in the metabolism of porphyrins are to be attributed to the activity of TCDD as a potent hepatic enzyme inducer.
6. The immunological survey highlighted changes that can clearly be correlated with exposure (increased complement and heightened *in vitro* response of lymphocytes to mitogens) but of uncertain meaning even though indicative of an altered immunological homeostasis.

The set of data at our disposal is, therefore, sufficient to suggest the possibility of a chronic, latent or subclinical pathology related to TCDD intoxication. As things stand at the moment it is impossible to reach a more definite conclusion....

The cytogenetic monitoring of selected samples of exposed individuals failed to indicate significant differences with the control group even though a tendency toward an increase was observed. Due to their low sensitivity these studies will probably be abandoned.

VI. NEUROLOGY

Neurological studies, among the clinical studies, deserve particular interest and have been the subjects of long debates, because their results were incomplete, but strongly suggestive. Neurotoxic effects in workers occupationally exposed to TCDD have been described, including: polyneuropathies, lower extremity weakness, and sensory impairment. On the other hand, properly controlled studies on this subject are still lacking.

Several neurological screenings have been carried out among residents of the Zone A from 1977. This included a questionnaire on medical history, a standard clinical examination, and an electrophysiological investigation. Subjects with alcoholism, with potentially neurotoxic occupational exposure, or with diseases predisposing to peripheral neuropathy were identified and analyzed separately. Reference "normal" values were calculated from a control population living in an unpolluted area, excluding again individuals belonging to the above-mentioned categories. Unfortunately, the screenings were not completely successful. Chloracne and raised serum liver enzyme levels were used as indicators of exposure.

Of the 277 individuals examined in 1978, clear association was established between electrophysiological diagnosis of neuropathy and presence of indicators of TCDD exposure.

The prevalence rate ratio (i.e., a relative risk) was 2 of 8, which was similar to that observed among those with known risk factors. The prevalence ratio was higher among those with raised liver enzyme levels than among those with chloracne. However, if only individuals under 20 years of age are considered, the prevalence rate ratio was even higher. These results are statistically significant, notwithstanding the small numbers involved. Thus, there is actual evidence in favor of the neurotoxic effects of TCDD in the Seveso area.

The main criticisms to these conclusions are concerned with the following:

1. The high rate of nonrespondents, which could have introduced some bias in selection.
2. The reliability of the electrophysiological parameters and of the ranges of normality.
3. The clinical relevance of the cases (most cases are subclinical or only instrumental).
4. The exiguity of the number of cases detected.

The debate on these issues is still open.

VII. OCCUPATIONAL MEDICINE

With regard to occupational medicine, health monitoring programs were started soon after the accident. These programs covered several groups, according to the Regional Law: former Icmesa workers (191 persons); former employees of factories located in Zones A and B; persons usually entering Zone A for particular tasks and/or reclamation; and military personnel employed in patrolling the fences of Zone A.

Due to the characteristics of these programs, and particularly to the lack of adequate study protocols, it is not possible to draw reliable conclusions from the large amount of data collected. However, 27 hospital admissions among the 654 decontamination workers of Zones B, A6, and A7 were reported. In six out of twenty-seven cases, a diagnosis of chronic liver disease in varying stages was made. Among the Icmesa workers (191 persons) from July, 1976 to December, 1979, a high incidence of various pathological conditions was observed; 33 individuals were admitted to hospitals with various medical problems. The high admission rate (33/191) indirectly indicates the overall poor health of these workers. No chloracne-type skin change was observed. In one-third (51) of the 154 workers who were examined yearly, pathological blood tests were observed at least once.

On September, 1981 a study group was established on Icmesa and decontamination workers. The research protocol is given below.

A. A cross-sectional study of the workers and subcontractors employed at the Icmesa plant on July 10, 1976 (191 people) and a control group. The control group was selected among workers employed on July 10, 1976 at two plants outside the polluted area, without any known exposure to hepatoneurotoxic agents (250 persons). This study was started June, 1981 and is on-going.

B. A prospective controlled study to measure the efficacy of the industrial hygienic procedures used during the clean-up of the highly polluted Zone A. In this study, an identical clinical follow-up was carried out on exposed workers and controls. The study was started on May, 1980. During the first 9 months of follow-up no remarkable changes were observed in a battery of laboratory tests between clean-up workers and exposed subgroups, who were classified according to the number of working hours spent in the Zone A.

C. A cause-specific mortality study among Icmesa workers from 1947 to 1976 is being conducted. The results will be compared with the mortality experience of the same control group employed in the morbidity study and with that of the general population of the 11 municipalities.

Preliminary results on 956 Icmesa employees should be available soon.

VIII. CHLORACNE

Particularly interesting is the geographical distribution of chloracne cases. More than twenty-nine cases were identified in the screenings.

Most chloracne cases are presently recovering, with residual disfiguring scars in a minority of cases. It is noteworthy that chloracne mainly affected children. This can be due either to the fact that screenings were carried out on school children, or that children were more frequently exposed.

Acute clinical findings among chloracne children include: urinary tract symptoms, inflamed joints, gastrointestinal symptoms, headache and eye irritation. Changes were noted in biochemical tests including GGT, GPT, and ALA-U.

New clinical studies on chloracne cases are presently being planned.

IX. CONCLUSIONS

It is very difficult to draw general conclusions on the health impact of the Icmesa accident on the Seveso population from this series of ambiguous, often incomplete reports. On the one hand, it is becoming clearer that the feared disaster has not taken place. Most of the positive results are of limited relevance. When dubious results are obtained, the effects are limited to few cases. On the other hand, it is fallacious to state that the Icmesa accident passed over the Seveso population with no effect other than the occurrence of chloracne. Several studies indicate the presence of early effects on liver, on blood lipids, on the peripheral nervous system, and on pregnancy.

More refined studies are needed to clarify these hypotheses, taking into account the presence, the conditions, and the length of exposure. As far as long-term effects are concerned, we all hope that these will never appear. Careful inquiry is necessary in order to state with confidence the future possible presence or absence of these effects.

DISCUSSION

DR. FRAWLEY: It can be concluded that there were no really serious adverse effects as a net result of the accident. Chloracne is the only obvious effect and these cases have been recovering with time. Obviously, it is hoped that later effects will not develop in these exposed people with time.

PROFESSOR TUCHMANN-DUPLESSIS: Birth defects are hard to relate to specific causes and it is even difficult to obtain experts who will agree on a diagnosis. Therefore, the data should be considered as still under study and it is important to continue the epidemiological work.

DR. COULSTON: Were there epidemiologic studies done on the malformations in children born to women exposed to dioxins at the time of the accident?

DR. BRUZZI: No one is sure if there were malformations that were caused by the accident, but the answer is no in those cases where there was an elected abortion following exposure. Cytologic studies were not revealing. It was agreed that, if the dioxins were potent carcinogens or teratogens, a more significant impact on the exposed population would have been obvious.

DR. COULSTON: Regardless of the nature of the abortion, nothing was found. In your opinion, is there any causal relationship between dioxin exposure and any malformation that showed up in these studies?

ISBN 0-12-193160-9

DR. *BRUZZI:* The data do exclude a major deviation from what was expected. The nature of the final effects is not known.

DR. *COULSTON:* If dioxin had been a real toxic agent in this situation more positive data would have been expected. In my laboratory high doses of dioxins have not produced any malformations in the many monkeys that were followed through their pregnancies. This finding certainly supports the epidemiologic studies at Seveso.

DR. *KORTE:* The emphasis placed on certain chemicals is both puzzling and surprising, despite the fact that the population is surrounded by thousands of them. Of these hundreds of thousands of chemical compounds, there is little or no knowledge of their true toxicity for man. Yet, for many reasons, often political, massive toxicologic and epidemiologic studies on a single substance are undertaken. A more rational attitude toward these so-called "disasters" is needed so that a proper public approach can be developed. If this is not done the toxicological resources of the world will not be utilized efficiently and the limited skills and talents may be exhausted by concentrating on less important toxicological issues. A major consideration to be resolved in the future is the proper extrapolation of the true hazards for man from animal studies that suggest toxicity based on the pathologic evidence in these animal species.

DR. *SHEPARD:* Health-related problems to TCDD are particularly important for those who are responsible for medical care of people exposed to potentially toxic substances. These problems can be divided into several groups: how the problems arose in Vietnam with Agent Orange; how the Veterans Administration became involved; and some general issues of toxicity.

CHAPTER 14

HEALTH-RELATED PROBLEMS TO TCDD*

Barclay M. Shepard

Veterans Administration
Central Office
Washington, D. C.

This chapter describes the nature of the military operations in Vietnam and why defoliants were employed. An overview of the methods used in the dissemination of the code-named substance, Agent Orange, which is a mixture of equal parts of 2,4-D and 2,4,5-T employed as a defoliant will be provided.

Agent Orange was reported to contain 1.98 ppm of TCDD as an industrial chemical contaminant. However, it is likely that the levels of contamination varied. One important fact from the standpoint of exposure of men to a potentially toxic substance, was the military use of the container drums of Agent Orange. These were employed in a very indiscriminant manner, because there was no reason to suspect that a hazard existed. They were, therefore, used as bunker protectors, as campsite

**This is an edited transcript; a complete manuscript was not submitted.*

ISBN 0-12-193160-9

containers, and in numerous other ingenious ways. Leakage of the contents was a common observation and many different individuals were required to transport the drums and transfer the herbicide to the disseminating aircraft. In all, many individuals were exposed to the herbicide in many ways and to various degrees.

As a result of the concern about possible toxic effects of dioxin contamination of Agent Orange the Veterans Administration (VA) has mounted a long-range epidemiological study of the possible health effects among Vietnam veterans exposed to this herbicide. Some heavily exposed groups were given special consideration, in terms of their post-Vietnam health, for any long-term effects. Areas of VA review include the methods of environmental transport and potentialities for delayed toxicity, prolonged chronic toxicity, carcinogenicity, and birth defects. In addition to the potential toxicity of 2,4-D, 2,4,5-T, and the contaminant 2,3,7,8-TCDD, other herbicides used in Vietnam are included in the VA study. In all this work it is obviously extremely difficult to determine the level of exposure for any specific individual. This will make toxicity evaluation almost impossible unless some specific biological marker is found; this does not seem likely.

The public alarm about the use of Agent Orange caused a VA response in terms of health examinations of concerned veterans, and the subsequent mounting of a major program within the Department of Medicine and Surgery to respond to these many and varied concerns. These activities continue at the present time and hopefully will lead to developing a public policy with respect to the most satisfactory manner of resolving toxicological issues of this character. The resolution of the problem will obviously take many years because no one knows what the long-term effects may be. As an illustration of the magnitude of the problem, the VA has now examined over 61,000 individuals for possible health effects related to Agent Orange exposure. The Center for Disease Control has also undertaken a special long-term study of the children born to Air Force men, with appropriate matched controls, who were suspected of being highly exposed to Agent Orange. The results will be available within the next few years. General mortality studies will be completed within 2 years. Morbidity studies will take a little longer and long-range studies are projected for the next 20 years.

DISCUSSION

DR. FRAWLEY: Dr. Shepard, when do you really expect these various VA activities to be completed?

DR. SHEPARD: The mortality studies will be completed this year. The moribidity studies should be complete about 6 months later. The prospective long-range studies will certainly take 20 years, depending on financial considerations.

DR. FRAWLEY: The social issue will be resolved long before that.

DR. CARR: You have an almost impossible problem to resolve. If it can be shown by an independent group such as this, that a substance like the dioxins is not truly hazardous in the amounts people are exposed to, is it conceivable that the social concepts can be reversed so that people will not insist on such freedom from hazard?

DR. SHEPARD: If the real worry can be resolved in the minds of the veterans, and if the Congress will agree, these costly and time-consuming activities may be relaxed. The scientific data that emerges from this meeting could add substantially to this resolution. Obviously, it is going to be extremely difficult to establish and document a no-effect level for the dioxins as it is for most substances.

DR. KEARNEY: Studies have been very carefully made on some agriculture workers handling and spraying phenoxy herbicides that relate to the safety of these substances. It

ISBN 0-12-193160-9

should be noted at the outset that the concentrations of herbicides used in Vietnam were from 4 to 10 times the amounts used in agriculture in the United States.

CHAPTER 15

PROBLEMS IN AGRICULTURE WITH TCDD

Philip C. Kearney

Pesticide Degradation Laboratory
Agricultural Research Service
U. S. Department of Agriculture
Beltsville, Maryland

I. INTRODUCTION

American agriculture is and has been one of the primary users of the phenoxy herbicides (2,4-dichlorophenoxy) acetic acid (2,4-D) and (2,4,5-trichlorophenoxy) acetic acid (2,4,5-T), and therefore has an interest in the safety of these herbicides. One estimate (Fowler and Mahan, 1972) put the United States production of 2,4,5-T for the 10-year period of 1960-1970 at about 106,310,000 lb (48,264,740 kg) on 7,939,000 acres (32,000 km^2) (Table I). A subsequent estimate by the USDA/States/EPA 2,4,5-T Assessment Team (1982) found that at least 10 million lb (4.5 million kg) of 2,4,5-T were used annually in the United States to control weeds and brush on lands used for timber, grazing, rights-of-way, and rice. Regardless of the exact use figures for 2,4,5-T, some segment of the agricultural community has been exposed to this compound for about 30 years.

The herbicide 2,4-D has been more widely used in American agriculture, and one estimate (Fowler and Mahan, 1972) showed

ISBN 0-12-193160-9

TABLE I. Production, Exports, and Producers' Domestic Disappearance of 2,4-D and 2,4,5-T (Acid Basis) in the United States, 1960-1970[a]

	Production		Exports[b]	Domestic disappearance[c]	
Year	*2,4-D*	*2,4,5-T*	*2,4-D & 2,4,5-T*	*2,4-D*	*2,4,5-T*
1960	36,185	6,337	8,796	31,131	5,859
1961	43,392	6,909	9,085	31,067	5,444
1962	42,997	8,369	10,192	35,903	8,102
1963	46,312	9,090	14,657	33,199	7,179
1964	53,714	11,434	13,037	43,986	8,912
1965	63,320	11,601	6,924	50,535	7,244
1966	68,182	15,489	5,419	63,903	17,080
1967	77,139	14,552	4,410	66,955	15,381
1968	79,263	17,530	3,391	68,404	15,804
1969	47,077	4,999	7,287	49,526	3,218
1970	43,576	--[d]	9,571	46,942	4,871

[a] *Data for 1000 lb (~454.5 kg).*

[b] *Excludes military shipments abroad; these are not considered exports.*

[c] *Includes military shipments abroad.*

[d] *Separate figure not available.*

that 601,157,000 lb (272,925,280 kg) were produced in the 10-year period 1960-1970. Based on actual farmers' use surveys, an estimated 40,420,000 lb (18,334,000 kg) were used in 1976. Currently, approximately 1500 2,4-D products are used for vegetation control in home lawns, forests, rights-of-way, drainage ditch banks, rangeland, pasture, aquatic areas, cereal crops, sugar cane, and commercial turf. Since 2,4-D was discovered and used prior to 2,4,5-T, a larger segment of the agricultural community has been exposed to 2,4-D for more than 35 years. In the present context, is that population at risk, and what steps have been taken to assess that risk? These questions are discussed in more detail in this chapter.

Before discussing some current studies on herbicide exposure and related health studies on farm and forestry populations exposed to the phenoxys, a brief review of some of the activities in the Department of Agriculture (USDA) merit consideration. When it was disclosed that 2,4,5-T was contaminated with the highly toxic 2,3,7,8-tetrachlorodibenzo-*p*-dioxin

(TCDD), USDA scientists undertook one of the first pesticide surveys to determine the amount and distribution of TCDD contamination (Woolson et al., 1972), and some of the first studies on soil persistence (Kearney et al., 1972), photodecomposition (Plimmer et al., 1973), mobility in soils (Helling, 1970), plant uptake (Isensee and Jones, 1971), kinetics in rats (Fries and Marrow, 1975), distribution and bioaccumulation in an aquatic ecosystem (Isensee and Jones, 1972), and in plants, soil, water, and air in a microagroecosystem (Nash and Beall, 1978). The USDA also maintains a current bibliography on all aspects of the phenoxy herbicide literature (Bovey and Young, 1980).

When questions arose about possible adverse health effects on Vietnam veterans exposed to Agent Orange (an equal mixture of the butyl esters of 2,4-D and 2,4,5-T), the USDA initiated or cooperated in several studies designed to learn more about exposure and possible health effects, if any, on the farm community from extended use of these two herbicides. It should be cautioned, however, that parallels cannot be drawn between the Vietnam experience and the use of these same two herbicides in United States agriculture for the following reasons:

1. Agent Orange has never been used commercially in American agriculture.
2. The rates of Agent Orange applications in Vietnam were 1.5 to 3 gallons per acre (1400 to 2800 l/km^2; or about 6 to 12 lb per acre or 672 to 1344 kg/km^2), as opposed to ½ to 1 lb per acre (0.56 to 1.12 kg/ha) for most uses of 2,4-D and 2,4,5-T in agriculture.
3. The mean TCDD concentrations of Agents Purple, Green and Pink, and Orange were 32.8, 65.6, and 1.98 ppm (Young et al., 1978), respectively, while the majority of domestic formulated samples of 2,4,5-T were less than 0.5 ppm, with only a small number greater than 10 ppm (Woolson et al., 1972).

Although these differences existed, the USDA nevertheless initiated or cooperated in several studies to learn more about the safety of these compounds. Issues raised by the Vietnam experience were birth defects and spontaneous abortions, cancer, and a third category of problems such as claims of loss of sex drive, fatigue, inability to concentrate, nervous disorders, liver problems, and related illnesses. The following studies were included:

1. An agricultural applicator exposure to 2,4-D and 2,4,5-T study conducted by the University of Arkansas and the Science and Education Administration, USDA.
2. A reproductive mortality and morbidity study on aerial applicators by the National Agricultural Aviation Association (NAAA).

3. A case control study of the relationship between exposure to 2,4-D and spontaneous abortions in humans sponsored by the National Forest Products Association and the Forest Service, USDA.

These studies are related since a segment of each investigation was in the same geographic region and on a group of people with similar occupations.

II. EXPOSURE STUDIES

Exposure measurements are extremely important in any risk equation since these measurements provide estimates of the actual dose received by a study population and the possible untoward effects to that population based on existing toxicology data. This toxicology data often comes from experiments on laboratory animals and, consequently, there are major problems in extrapolating this data to man, as has been discussed by Squire (1981). Nevertheless, toxicology is the basis for many decisions on the continued registration of many pesticides. Exposure data then bridges the gap between toxicological data and an assessment of risk to man.

For an excellent review of exposure studies with the phenoxy herbicides in humans, including applicators under field conditions, the reader is referred to an article by Leng et al. (1982). There are a number of ways of measuring exposure to pesticides. Direct measurements of exposure are usually based on chemical analysis of air samples and of patches simulating skin. These patches are placed at strategic positions on the clothing of workers and later analyzed for residues. It is important, however, to distinguish exposure measurements and dose. Indirect measurements of actual dose are based on chemical analyses of blood and urine samples taken periodically after the application. A third, and probably least satisfactory method, is from environmental samples taken from the spray area vicinity and back extrapolation from decay curves to estimate the amount applied and probable exposure to people in that area. Although crude, this method may offer the only approach to estimating exposure from some historic event, such as the Vietnam experience.

Laboratory research with human volunteers showed that only about 6% of the 2,4-D applied to the forearm was recovered in urine collected over 1 week after exposure (Feldman and Maibach, 1974). In a more recent study with a limited number of participants, less than 0.5% of the 2,4,5-T applied to the skin in the upper thigh region was measured in urine collected over a 5-day exposure (Newton and Morris, 1981). For dermal exposure, the rate-controlling step was shown to be adsorption of the water through the skin (Leng et al., 1981). Once in the body, these esters are rapidly hydrolyzed to the acid, and the acid rapidly excreted.

III. ARKANSAS STUDIES

These studies were conducted by Lavy (1979, 1980) using 2,4,5-T under field conditions for forestry applications. A total of 21 participants using four types of equipment, including backpack sprayers, a mistblower, and two helicopter delivery systems were examined. Normal clothing and routine operating procedures were employed by the four crews under study. The study demonstrated the variability of exposure of the 21 participants, depending on the job performed and on the care taken by the individual to avoid exposure. Table II summarizes the total 2,4,5-T excreted as a function of job description by the various spray crews.

IV. USDA STUDIES

To determine exposure levels in agricultural applicators to 2,4,-D applied for weed control in wheat under normal use conditions, residues in urine samples were measured (Nash et al., 1982). Participants included 26 ground or tractor spray applicators in eastern North Dakota after a single exposure and 17 aerial applicators in eastern Washington during intermittent exposure. Prior to the experiments, each participant completed an information sheet which listed age, sex, weight, job, years of exposure, and type of applicator equipment. Information on clothing and weather conditions were collected during the experiments.

Several important observations were gleaned from these experiments. As in the Lavy study, the job of a participant strongly influenced exposure. For example, participants that were responsible for mixing and loading the herbicide for aerial application received a higher dose (0.02 mg/kg mean daily urinary excretion based on body weight) when compared to pilots (0.006 mg/kg)--roughly 3 times more. For ground applicators, the maximum mean 1 day 2,4-D urinary excretion was 0.002, 0.002, and 0.004 mg/kg for applicator, mixer/loaders, and mixer/loader/applicators, respectively, for a one-time exposure. A correlation was found to exist between 2,4-D exposure and job (e.g., tractor driver, pilot, loader/mixer) for both aerial and ground applicators. In addition, a correlation existed between 2,4-D excreted in urine versus hours of exposure and versus amount (lb) of 2,4-D applied by ground applicators.

TABLE II. Summary of Total 2,4,5-T Measured in Urine of Spray Crews per Exposure under Field Conditions in Forests in Arkansas[a]

Job description	*Number of subjects*	*Range of mg 2,4,5-T per kg body wt*[b]
Mixers	4	0.012 - 0.138
Backpack sprayers	6	0.019 - 0.104
Spray tractor drivers	2	0.033 - 0.049
Helicopter pilots	2	< 0.001 - 0.044
Supervisors	3	0.002 - 0.030
Flagmen	4	< 0.001 - 0.003

[a]*Adapted from Lavy.*

[b]*Includes 2,4,5-T excreted on day 0 before each spraying due to wearing contaminated clothing or exposure of fingers during mixing of spray, and 2,4,5-T excreted on day 6 after first spraying (i.e., on day 0 of second spraying).*

Some conclusions from the Lavy exposure study on forestry workers as summarized by Leng (1982), and our own work are as follows:

1. The major route of exposure to 2,4-D and 2,4,5-T leading to a measured dose by urine residue analyses, is via skin absorption and not inhalation.
2. In the worst case situation where the worker does not use protective clothing and appropriate handling precautions, the maximum expected exposure under field conditions for 2,4-D and 2,4,5-T would be less than 0.1 mg/kg body weight per day. Normal exposure levels are usually less than one-half that value or <0.05 mg/kg body weight.
3. Highest exposure occurs in the mixer and/or loader who handled the herbicide concentrate. This exposure could be considerably reduced if workers wore gloves. Applicators using backpack sprayers, spray tractor drivers, pilots, and flagmen, in approximately that order, receive considerably lower doses.
4. The amount of herbicide received by a worker is directly proportional to the time or number of hours exposed to the chemical and the amount or weight in pounds of that chemical handled in any spray operation.
5. Measurement of contamination in patches worn by field workers is of limited value in estimating actual exposure to herbicides such as 2,4,5-T.

6. Bystanders or persons entering treated areas after spraying are not likely to receive measurable exposure.

7. Applicators whose lifetime occupation has been farming have received low dosages. The median work experience for the North Dakota wheat farmer was 10 years, with a range of 1 to 31 years.

V. NAAA STUDY ON REPRODUCTIVE MORTALITY AND MORBIDITY

While the exposure studies were in progress, the membership of NAAA funded an investigation on the possible effects of pesticide exposure on reproductive mortality and morbidity. We were interested in this study since some of the same pilots in our 2,4-D exposure study (Nash et al., 1981) were included in the NAAA survey, and, consequently, baseline exposure data existed for an occupational group undergoing a study for miscarriages and stillbirths after male exposure to 2,4-D. A preliminary report (Roan, 1980) on comparisons between populations of agricultural pilots and their siblings who are not occupationally exposed to pesticides has appeared. A total of 653 people participated in the study comprising 374 from the agricultural aviators (196 males and 178 females) and 279 from the siblings (136 males and 143 females). This population was fairly evenly distributed around the United States. A total of 373 live births with 28 (8%) miscarriages and stillbirths were recorded for the aerial applicators, and 360 live births with 47 (12%) miscarriages and stillbirths were recorded for their siblings. Thus, no apparent correlation was established between exposure to pesticides and reproductive mortality. The limited number of participants was a major problem in this study. Nevertheless, with this limited but select population, no adverse reproductive effects were observed.

VI. CASE CONTROL STUDY ON 2,4-D EXPOSURE AND SPONTANEOUS ABORTIONS

In May 1980, the U. S. Forest Service cosponsored a case control study of the relationship between exposure to 2,4-D and spontaneous abortions in humans with the National Forest Products Association through a contract to SRI International. The study population included grain farmers, forest workers, and herbicide applicators in the states of Oregon and Washington. Again, part of our exposure study (Nash et al., 1981) merged with the SRI study in that exposure data existed for a segment of the population under examination in the same geographic and occupational areas.

The epidemiological design of the spontaneous abortions survey was a retrospective case control study. Results of the 1 year-long study on male farm, forest, and commercial workers who were occupationally exposed to 2,4-D were recently reported (Carmelli et al., 1981). Miscarriage rates for wives of men exposed to 2,4-D were compared to miscarriage rates for wives of men who were not exposed. The data base included 134 miscarriages (cases) and 311 live births (controls). Considerable effort was expended in documenting both the occurrence of spontaneous abortions and occupational exposure. Analysis of the data revealed no evident relationship between use of 2,4-D and spontaneous abortion. One of the analyses, however, produced an inconsistent result in which a small subgroup of young forest and commercial workers exhibited a "suggestive association" with 2,4,-D exposure. No association was observed for the same group of farmers who could be expected to have an equal or greater exposure to 2,4-D. This study again included a small number of participants, however, only a limited number of people share the unique requirements for such an epidemiological survey.

VII. SUMMARY

The subject of this volume deals with the human health aspects to accidental chemical exposure to dioxins. In a sense, the agricultural community, through extensive use of 2,4-D and 2,4,5-T, has been exposed to a very low level of various dioxin-containing herbicides for about 30 years. When questions arose about the level of exposure to these herbicides and the possibility of miscarriages or stillbirths due to exposure, several studies were undertaken. Analytically it was possible to document exposure to 2,4-D in two typical work situations in agriculture, i.e., ground and aerial application. In the same time frame, two health studies were conducted on worker exposure to 2,4-D and spontaneous abortions. The failure to detect any association between 2,4-D usage and spontaneous abortions in a population subjected to continuous, albeit, very low level exposure is extremely encouraging. The two health studies taken individually provide limited information on the effects of the phenoxy herbicides on public health. They do, however, make a contribution to a growing body of scientific evidence that no correlation exists between exposure to these herbicides and miscarriages and stillbirths. They also serve as a basis for addressing situations where potentially larger exposures have occurred, but where that exposure cannot be fully documented.

REFERENCES

Bovey, R. W., and Young, A. L. (1980). "The Science of 2,4.5-T and Associated Phenoxy Herbicides," John Wiley and Sons, New York, 462 pp.

Carmelli, D., Hofherr, L., Thomsik, J., and Morgan, R. W. (1981). Final Report, A Case Control Study of the Relationship Between Exposure to 2,4-D and Spontaneous Abortions in Humans, prepared for the National Forest Products Assoc. and U.S.D.A. Forest Service, Washington, D. C. 128 pp.

Feldmann, R. J., and Maiback, H. I. (1974). *Toxicol. Appl. Pharmacol. 28,* 126-132.

Fowler, D. L., and Mahan, J. N. (1972). The Pesticide Review 1971, Agricultural Stabilization and Conservation Service, U. S. Department of Agriculture, Washington, D. C.

Fries, G. F., and Marrow, G. S. (1975). *J. Agric. Food Chem. 23,* 265-269.

Helling, C. S. (1970). *Soil Sci. Soc. Amer. Proc. 35,* 737-743.

Isensee, A. R., and Jones, G. E. (1971). *J. Agric. Food Chem. 19,* 1210-1214.

Isensee, A. R., and Jones, G. E. (1975). *J. Agric. Food Chem. 9,* 668-672.

Kearney, P. C., Woolson, E. A., and Ellington, C. P. (1972). *Environ. Sci. Technol. 6,* 1017-1019.

Lavy, T. L., Shepard, J. S., and Mattice, J. D. (1980). *J. Agric. Food Chem. 28,* 626-630.

Lavy, T. L., Shepard, J. S., and Bouchard, D. C. (1980). *Bull. Environ. Toxicol. 24,* 90-96.

Leng, M. L., Ramsey, J. C., Brun, W. H., and Lavy, T. L. (1982). *In* "Pesticide Residues and Exposure," (J. R. Plimmer, ed.). American Chemical Society, Washington, D. C. Symposium Series 182, 133-156.

Nash, R. G., and Beall, M. L., Jr. (1980). *J. Agric. Food Chem. 28,* 614-623.

Nash, R. G., Kearney, P. C., Maitlen, J. C., and Sell, C. R., (1982). *In* "Pesticide Residues and Exposure", (J. R. Plimmer, ed.). American Chemical Society, Washington, D. C. Symposium Series 182, 119-132.

Newton, M., and Norris, L. A. (1981). "Fundamental and Applied Toxicology" 1, 339-346.

Plimmer, J. R., Klingebriel, U. I., Crosby, D. C., and Wong, A. S. (1973). *In* "Chlorodioxins: Origins and Fate" (E. H. Blair, ed.). American Chemical Society, Washington, D. C. Adv. Chem. Ser. 120, 49-52.

Roan, C. C. (1980). Part I. "Preliminary report comparisons between populations of agricultural pilots and their siblings who are not occupationally exposed to pesticides." *World of Agric. Aviation 7,* 12-34.

Squire, R. A. (1981). *In* "The Pesticide Chemist and Modern Toxicology" (S. K. Bandal, G. J., Marco, L. Goldberg, M. L. Leng, eds.). American Chemical Society, Washington D. C. Symposium Series 160, 493-501.

U. S. Dept. Agric./State/EPA (1982). Coop. Impact Assessment Report. The Biological and Economic Assessment of 2,4,5-T. USDA Tech. Bull. 1671. 475 pp.

Woolson, E. A., Thomas, R. F., Ensor, P. (1972). *J. Agric. Food Chem. 20,* 351-354.

Young, A. L., Calcagni, J. A., Thalken, C. E., and Tremblay, J. W. (1978). The Toxicology, Environmental Fate and Human Risk of Herbicide Orange and Its Associate Dioxin, USAF. OEHL Technical Report, Brooks Air Force Base, Texas. 400 pp.

DISCUSSION

DR. BRAUN: A large study on 2,4-D was carried out in Oregon and Washington States under different exposure conditions where people dress more warmly and, hence, are not exposed physically to the degree found in warmer climates, such as Arkansas. In general, in this study, the levels of exposures were significantly less and in the order of a magnitude less. Hand contact is now known to be important as a source of body exposure to the herbicide. In the Washington and Oregon studies, gloves were routinely worn and there was a resulting decrease in the dosage absorbed as determined by measured urinary levels. This work will soon be published.

DR. KORTE: There should be some public agreement on the issue of assessing the relative toxicity of natural versus synthetic chemicals. Natural substances are generally more toxic. Yet there seems to be relatively little concern about these substances which are found everywhere in man's environment. It seems that for various reasons emotional issues are recognized and discussed far beyond their level of true significance. The chemical industry has to constantly defend itself against charges that their products are causing all kinds of strange toxicities. Natural products are seldom brought into this arena.

DR. COULSTON: When concerned citizens write letters to regulatory agencies and the Congress about toxicity issues, the regulatory agencies are prodded into premature action. Actions taken in such an environment and under such extreme pressure can seldom be based on true scientific data.

ISBN 0-12-193160-9

This is especially true when the facts are not known. This is not to say that proper regulatory response to correct and recognize scientific toxicological hazards is not in order. On the other hand, over-reaction to emotional concerns is not the way to resolve public issues when, given time and resources, the true facts can be learned. In the case of dioxins it is becoming clear, as has been stressed that the hazards are extremely low. We are now confronted with the proposition of what is to be done when the next accident happens. There must be a mechanism to respond to honest public concern. This we have. There must also be a plan to guard against over-reaction. Our future decisions are at stake at this time and our present concerns illustrate the need for adequate planning for the future.

DR. FRAWLEY: There is no question that when people over-react, the emotional trauma created is a form of injury that indeed may be greater than the toxicity we are trying to control.

CHAPTER 16

USE OF EPIDEMIOLOGY IN THE REGULATION OF DIOXINS IN THE FOOD SUPPLY

Frank Cordle

Epidemiology & Clinical Toxicology
Bureau of Foods
Food & Drug Administration
Washington, D. C.

I. INTRODUCTION

Although epidemiology may have first been used by Hippocrates, the definition or meaning has undergone some significant changes since then. First described as the study of epidemics and, later, as the study of the determinants of differences in disease distributions in human populations, epidemiology today, especially in regulatory agencies, refers essentially to the activities of the epidemiologist. In a regulatory agency such as the Food and Drug Administration (FDA) the role of the epidemiologist is to assimilate, digest, and synthesize the best available data from survey, clinical, and laboratory experiences, either firsthand or from the scientific literature; these human and animal data are then combined and interpreted in order to address a basic question: Does the evidence of adverse human health effects from exposure to the various chemical substances that fall under FDA regulatory authority present a risk to public health safety that is sufficient for

ISBN 0-12-193160-9

FDA to initiate or make a change in regulatory action? The importance of the contribution of epidemiology in the regulatory public health arena seems obvious. Without adequate assessment of factors that appear to increase the risk of disease in humans from chemical substances in the food supply, food may be destroyed, the public may be unnecessarily inconvenienced or inadequately protected, and the costs may be enormous. In short, epidemiological efforts are directed toward the prediction of such risk at the community, state, or national level.

It is quite likely that the majority of humans are exposed to a large number of chemical substances in small amounts over an extended period of time rather than to large doses, such as those described in the mercury episode in Minamata or in the PCB exposure in Japan. The importance of the possible cumulative effects from these small doses versus the importance of the effects from a single large dose or from relatively large doses over a short period of time is the subject of considerable scientific debate.

Much of the information concerning the toxicity of various chemical substances has been obtained from experiments in animals. In general, safety testing depends upon the fact that as the exposure dose is decreased, toxic effects also decrease, and a dose is finally established at which "no observed effects" are seen. Such dose-response relations are central to toxicity studies in animals and should be of equal concern in the evaluation of human exposure to a variety of environmental and other insults. However, dose-response relations in humans must be established or identified with a great deal of care and caution. Outcome may vary greatly in significance: At one extreme is death and at the other is change in physiological or psychological function. The weight that should be given to each relation in formulating regulatory policy may vary greatly.

The lack of good examples of dose-response relation, even in the area of occupational exposure, is remarkable. Where dramatic incidents of exposure to environmental contamination have occurred through ingestion of contaminated food, efforts to establish an acceptable dose-response relation in humans have met with little success.

However, epidemiological study should not be limited to outbreaks of disease that are obviously caused by high exposures to toxic chemicals; it should be equally concerned with the effects of lower, more prolonged, and sometimes insidious exposures.

II. EPIDEMIOLOGICAL METHODS

In the traditional manner, epidemiologists deal with epidemics by interpreting patterns of disease, testing hypotheses, and assessing the risks and benefits of various options. Their methods involve the use of two categories of epidemiology often referred to as analytical and descriptive. The analytical epidemiological methods generally involve either a case-control, retrospective approach, or a cohort, prospective approach.

In analytical epidemiology, the goal of epidemiological activity is to identify and, if possible, to quantify the association between a causal exposure or characteristic and a disease under circumstances that permit the best possible discrimination between cause and alternative hypotheses. Chance will always affect the alternatives; thus the feasibility of such a study is limited by factors such as sample size. Since a precise hypothesis permits the use of a well-thought-out and properly planned study design, it is better to use the cohort approach to measure relative and attributable risk, for example, than to try to detect trends from descriptive data. Nevertheless, sample size considerations, cost, and time constraints often restrict the kind of epidemiological study undertaken, even though the cohort approach might be preferred for a given problem.

Case-control studies of disease, i.e., studies looking at the past history of exposure, are often more feasible than cohort studies because they can be conducted in a relatively short time and are easily repeatable, and because a large number of cases can be studied economically. They do have some special liabilities, however, in terms of validity, since such studies depend entirely on the comparability of cases and controls and on the specific methods used to measure the exposure. Although frequently used, the methods for estimating relative risk or risk/ratio from case-control studies also present some special problems of inference.

Cohort studies, i.e., studies looking forward in time, are especially costly because of the time involved in the prospective follow-up. They do, however, provide an opportunity to compute a more reasonable relative and attributable risk, and specious relationships resulting from bias in data collection are less likely to occur.

Descriptive epidemiological methods can be used by a regulatory agency such as FDA to determine trends in disease and magnitudes of exposed populations largely from data that are readily available, e.g., mortality data, data from National Cancer Institute (NCI) surveys, census data, and food consumption data. The goal of these studies is to identify any unexpected changes in incidents or mortality through surveillance of the available data for time trends and through probes of specific data collected during research activities or gathered in support of regulatory decisions.

Epidemiological studies also provide information that is used to identify and quantify differences in species responsiveness to environmental agents in a variety of other ways. One use, for example, pertains to the difficulty associated with comparing data derived from studies of highly inbred animal strains to data of a genetically heterogeneous human population. In this case, epidemiological methodologies provide the means for stratification of study data according to sex, age, race, and other variables that characterize human populations. Certain problems may also be minimized when epidemiological considerations are employed in risk assessments that are based on comparisons involving data of humans and data of animal models. These include problems associated with limited sample sizes, migration in and out of the exposure area, and toxicity due to causes other than that associated with exposure to the etiologic agent under study.

III. EPIDEMIOLOGICAL AND TOXICOLOGICAL ASSESSMENT OF TCDD EXPOSURE

Recent concern for the potential human health risk associated with exposure to residues of the tetrachlorodibenzo-*p*-dioxins (TCDD) in several species of freshwater fish from various areas of the Great Lakes has resulted in an epidemiological and toxicological assessment of the problem that utilizes currently available data from a variety of data sets. For a regulatory decision to be made, several elements were needed. These included data for the TCDD residues in the freshwater fish, for the amounts of the various fish species consumed, and for the number of individuals who consumed the fish, and assessments of pertinent animal data and previous human exposure to TCDD.

A. ANIMAL DATA

A number of toxicological studies with TCDD have been conducted to assess the potential for acute toxicity and teratogenesis. Kociba et al. (1976) reported the results of a subchronic study using rats that were given 1.0, 0.1, 0.01, 0.001, or 0 μg TCDD/kg body wt/day for 5 days/week for 13 weeks. Doses of 1.0 μg TCDD/kg/day caused some mortality, inactivity, decreased body weights and food consumption, pathomorphologic changes in the liver, lymphoid depletion of the thymus, increased urinary excretion of porphyrins, and minimal alterations of some hematopoietic components. Doses of 0.1 µg TCDD/kg/day caused decreased body weights and food consumption and slight degrees of liver degeneration and lymphoid depletion. These data indicate that no discernible ill effects occurred in rats given 0.01 or 0.001 µg TCDD/kg/day for 5 days/week for 13 weeks.

In a 2-year chronic study in rats, Kociba et al. (1978) reported that the ingestion of 0.1 µg/TCDD/kg/day caused an increased incidence of hepatocellular carcinomas and squamous cell carcinomas of the lung, hard palate/nasal turbinates, or tongue, whereas a reduced incidence of tumors of the pituitary, uterus, mammary glands, pancreas, and adrenal glands was noted. Other indications of toxicity at this dose level included increased mortality, decreased weight gain, slight depression of erythroid parameters, and increased urinary excretion of porphyrins. Gross and histopathologic changes were noted in the hepatic lymphoid, respiratory, and vascular tissues. The primary heptatic ultrastructural change at this high dose level was proliferation of the rough endoplasmic reticulum. Terminal liver and fat samples from rats given this high dose level contained 24,000 and 8100 parts per trillion (ppt) TCDD, respectively. Rats given 0.01 µg TCDD/kg/day for 2 years showed less severe toxicological effects than those given the highest dose level. These included liver lesions (including hepatocellular nodules) and lung lesions (including focal alveolar hyperplasia). Terminal liver and fat samples from rats given this dose level contained 5100 and 1700 ppt TCDD, respectively. Ingestion of 0.001 µg TCDD/kg/day (22 ppt in the diet) caused no effects of any toxicological significance. Terminal liver and fat samples from rats given this low dose level each contained 540 ppt TCDD.

In a 3-year reproduction study in rats given a dose level of 0, 0.001, 0.01, or 0.1 µg TCDD/kg/day, Murray et al. (1979) reported no significant toxic effects in the F_0 rats of either sex during the 90 days of TCDD ingestion prior to mating. However, significant decreases in fertility and neonatal survival were observed in the F_0 generation rats receiving 0.1 µg TCCD/kg/day. At 0.01 µg TCDD/kg/day, fertility was significantly decreased in the F_1 and F_2 generations, but not in the F_0 generation, and decreases in litter size at birth, gestational survival, and neonatal survival and growth were also noted. Among the rats receiving 0.001 µg TCDD/kg/day, no effect on fertility, litter size at birth, or postnatal body weight was observed in any generation. No consistent effect on neonatal survival was observed at a dose level of 0.001 µg TCDD/kg/day.

Allen et al. (1977) reported results of a subchronic study in which female rhesus monkeys consumed a diet containing 500 ppt TCDD for periods as long as 9 months. It was calculated that these monkeys ingested a total of 2-3 μg TCDD/kg/body wt over the course of the 9-month study. Clinically, these monkeys showed changes similar to those described by McConnell et al. (1978), as well as some hematologic depression and hemorrhages in various tissues. Hypertrophy, hyperplasia, and/or metaplasia were noted in the epithelium of the bile ducts, salivary glands, bronchi, pancreatic ducts, sebaceous glands, skin, gastric linings, and urinary tracts of these monkeys given diets containing 500 ppt TCDD.

B. *HUMAN EXPERIENCE*

Although 22 incidents of human exposure to TCDD in connection with the manufacture of chlorinated phenols have been reported world-wide since 1949 (Holmstedt, 1980), there remains a scarcity of reliable information concerning the results of these exposures.

In a recent report of the mortality experience of a cohort of workers exposed to TCDD in Nitro, West Virginia in 1949, Zack and Suskind (1980) described some of the signs and symptoms observed in the exposed population shortly after the accident occurred. Employees who worked in the area of 2,4,5-trichlorophenol (TCP) production or were involved in the clean-up began to develop symptoms immediately following exposure to the material, which was discharged from the autoclave. Symptoms included eye and respiratory tract irritation, headache, dizziness and nausea, and a severe irritant reaction of the exposed skin. After these initial symptoms subsided, chloracne and other symptoms became evident. A total of 12 more severely affected workers were examined on three occasions during the period 1949-1953. Another 26 persons with chloracne that was apparently unrelated to the accident were also examined in 1953.

The clinical symptoms included acneform lesions; severe pains in muscles of upper and lower extremities, shoulders, and thorax on exertion; fatigue; nervousness and irritability; decrease in libido; dyspnea; vertigo; and intolerance to cold. All of the cases showed evidence of chloracne. For the six workers examined during 1949 and 1950, another examination was carried out in 1953, and at that time six additional workers involved in the accident were also examined. The findings in this later examination indicated a general regression of both the cutaneous and noncutaneous symptoms which had been present earlier. No specific levels of exposure could be determined.

In other reports of industrial exposure to TCDD from Great Britain, the Netherlands, West Germany, and Czechoslovakia, chloracne was the most common and prominent sign observed following exposure. In some reports liver function tests indicated liver damage, whereas in other reports they did not. Two major problems encountered in all of these studies were the lack of a clear identification of those exposed, other than their subsequent development of chloracne, and the absence of any measures for the levels of exposure that might have taken place.

Pazderova-Vejlupková et al. (1981) reported results of a 10-year follow-up study of workers exposed to TCDD between 1965 and 1968 during the production of 2,4,5-trichlorophenoxyacetic acid (2,4,5-T). In this study group of 55 individuals (originally 80 of the 400 persons engaged in the production became ill), the first indications of illness were feelings of sickness, fatigue, weakness in the lower extremities, and the formation of chloracne. Subsequent examinations indicated that about 20% had mild hepatic lesions.

During the 10-year follow-up study of these exposed individuals, most of the patients did not experience all of the symptoms and signs of intoxication, and some patients showed the same symptoms and signs as others, but in different combinations. It is assumed that in this type of intoxication all the systems and organs mentioned in the study were simultaneously affected, although some were affected only slightly. This assumption is supported by several facts. Fluorescence of liver tissues in uv light, which is a sign of pathological porphyrin metabolism, was present in all cases of necropsy and biopsy, i.e., in persons for whom long-term monitoring of porphyrin excretion in urine was carried out and in whom α -aminolevulinic acid values were constantly within normal limits. Probably a slight subclinical lesion was present in each of these patients. Further evidence was furnished in repeated neurological examinations. Polyneuropathy of the lower extremities was manifest in some patients only in the third or fourth year of illness. There is definite clinical and electromyographic evidence that the results of the first examinations conducted when the illness commenced showed that the patients were entirely normal.

Additional study results have been reported recently by Zack and Suskind (1980), Ott et al. (1980), and Cook et al. (1980), describing the mortality of employees engaged in the manufacture of 2,4,5-T. In two of these studies, cohorts of employees were assembled on the basis of their exposure to TCDD, which was indicated by the presence of chloracne; the third cohort consisted of individuals employed over the same time period. Unfortunately, each of these three cohorts was composed of a limited number of individuals; e.g., the chloracne groups contained 121 and 49 individuals and the employee group contained 204.

In each of these studies there does not appear to be an apparent excess in total mortality rate or in deaths from malignant neoplasms. It must be pointed out, however, that each of these studies does have limitations both in the size of the population studied and in other methodological areas, such as exposure levels.

For a more detailed description of these studies as well as the Seveso incident, the reader is referred to the original papers and to Homstedt (1980), to Reggiani (1980), and to Pocchiari et al. (1979).

C. *HUMAN EXPOSURE TO TCDD RESIDUES IN FISH*

Although there are no epidemiological studies which have firmly established an association between cancer in humans and TCDD exposure, and although comparisons of human exposure data and animal studies appear to indicate that humans may be less sensitive than animals to TCDD exposure, prudence dictates that human exposure to TCDD be kept to a minimum. With this public health objective in mind, the FDA recently completed an assessment of the problem of TCDD residues in some species of freshwater fish in the Great Lakes for the purpose of determining whether consumption of such fish provided a potential public health problem.

Results of the analyses of fish samples collected in Canada and the United States indicated that TCDD levels as high as 30 ppt with an average value of 25 ppt were present in the edible portion of salmonoid fish (salmon, trout) from Lake Ontario. Lower levels of TCDD were reported to be present in the edible portions of commercial species (bullhead, perch, catfish, sucker, etc.) from Lake Ontario, although up to 40 ppt TCDD were found in eel and smelt from the lake. Less than 10 ppt TCDD was seen in samples from Lake Erie, and limited data for fish from the other Great Lakes were similar to those obtained from Lake Erie fish, except that higher levels of TCDD were present in fish from Saginaw Bay, Michigan.

Data on fish consumption from the National Marine Fisheries Service (NMFS) were extracted for the eight Great Lakes states, Illinois, Indiana, Michigan, Minnesota, New York, Ohio, Pennsylvania, and Wisconsin. These data were collected in a national sample of households, consisting of approximately 25,000 individuals. From the data presented, individuals who consumed the species of interest, i.e., the species currently being analyzed for dioxin residues, were identified in the total sample, and the mean fish consumption in grams per day was computed. The 90th and 99th percentiles were also computed. On the basis of the proportion of individuals in the sample population consuming the species of interest, 17,057,791 individuals in the eight Great Lakes states would be expected to consume these same species. It must be pointed out that the number of consumers in the total population can be considered only an estimate because of the lack of information on potential sampling error as well as other factors. These data indicated that the 99th percentile for daily consumption for all consumers of the selected species was 36.8 gm; the 90th percentile for daily consumption was 15.7 gm. For a single species of fish such as pike, which appeared to be consumed in larger amounts than all the species combined, the 99th percentile of consumption was 83.95 gm. Table I, which is based on these fish consumption data, shows the daily human consumption for a range of hypothetical dioxin residue levels.

Thus, the four elements of the model for a scientific regulatory assessment, i.e., animal data, human data, consumption data, and residue data, were available. In this instance, the human data concerning TCDD exposure contributed little to the regulatory decision because of the uncertainty of the exposure data as well as the uncertainty of the outcome of the exposure. It was necessary, therefore, to rely on extrapolation from animal to human. In this case the rodent data from the 2-year chronic feeding study (Kociba et al., 1978) were used for the extrapolation.

Results of that study showed that (a) at 0.001 μg/kg body wt/day, no adverse effects were noted in rats exposed for their lifetimes to the dioxin; (b) at 0.01 μg/kg body wt/day, hyperplasia of the liver and lung was observed to occur, thus indicating an observable effect related to enzyme induction and liver cell response to the compound; (c) at 0.1 μg/kg body wt/day, an increase in liver carcinomas was observed.

The animal-to-human extrapolation of the no-effect level for TCDD exposure from the rodent data indicated an intake of 1 ng/kg body wt/day or a total daily intake of 70 ng as the no-effect level. If fish containing average TCDD residue levels of 25 ppt were consumed in the amount of the 99th percentile, i.e., 36.8 gm/day, the total daily intake of TCDD would be 0.92 ng or 13 pg/kg body wt/day or less than 1/70th of the no-effect level. If fish containing average TCDD residue levels of 25 ppt were consumed in the amount of the 99th percenmargin at the 90th percentile of consumption is even greater.

As a result of this assessment, a public health advisory was sent to the health officers in each of the Great Lakes states, encouraging a continued monitoring of TCDD residues in fish, especially in certain local areas where TCDD levels might be higher than the average, and where local fish consumption

TABLE I. Human Consumption of Dioxin Based on Fish Consumption Data

Hypothetical dioxin residue level in fish consumed (ppt[a])	*Total daily intake (ng[b])*			*Total daily intake/body wt (pg[c]/kg)*		
	All selected species		*Pike*	*All selected species*		*Pike*
	90th percentile	*99th percentile*	*99th percentile*	*90th percentile*	*99th percentile*	*99th percentile*
100	1.57	3.683	8.4	22	50	120
50	0.77	1.84	4.2	11	26	60
25	0.38	0.92	2.1	5.5	13	30

[a] *ppt = parts per trillion.*

[b] *micrograms (μg) = 1000 nanograms (ng).*

[c] *ng = 1000 picograms (pg).*

might also be higher than that for the Great Lakes area as a whole.

This example in regulatory decision-making illustrates the FDA's role in protecting public health by ensuring a safe and nutritious food supply. This responsibility is exercised by individuals who call upon science, law, and the regulatory process to accommodate the demands of safety, contamination, food requirements, Congress, consumers, and the courts. Although the regulatory process has often been criticized, on the whole it has provided effective public health protection for the millions who consume food in the United States.

REFERENCES

Allen, J. R., Barsotti, D. A., Van Miller, J. P., Abrahamson, L. J., and Lalich, J. J. (1977). Morphological changes in monkeys consuming a diet containing low levels of 2,3,7,8-tetrachlorodibenzo-*p*-dioxin. *Food Cosmet. Toxicol. 15,* 401-410.

Cook, R. R., Townsend, J. C., Ott, M. G., and Silverstein, L. B. (1980). Mortality experience of employees exposed to 2,3,7,8-tetrachlorodibenzo-*p*-dioxin. *JOM J. Occup. Med. 22,* 530-532.

Holmstedt, B. (1980). Prolegomena to Seveso. *Arch. Toxicol. 44,* 211-230.

Kociba, R. J., Keeler, P. A., Park, C. N., and Gehring, P. J. (1976). 2,3,7,8-Tetrachlorodibenzo-*p*-dioxin (TCDD): Results of a 13-week oral toxicity study in rats. *Toxicol. Appl. Pharmacol. 35,* 553-574.

Kociba, R. J. et al. (1978). Results of a two-year chronic toxicity and oncogenicity study of 2,3,7,8-tetrachlorodibenzo-*p*-dioxin in rats. *Toxicol. Appl. Pharmacol. 46,* 279-303.

McConnell, E. E., Moore, J. A., and Dalgard, D. W. (1978). Toxicity of 2,3,7,8-tetrachlorodibenzo-*p*-dioxin (TCDD) in rhesus monkeys *(Macaca mulatta)* following a single oral dose. *Toxicol. Appl. Pharmacol. 43,* 175-187.

Murray, F. J., Smith, F. A., Nitschke, K. D., Hameston, C. G., Kociba, R. J., and Schwetz, B. A. (1979). Three-generation reproduction study of rats given 2,3,7,8-tetrachlorodibenzo-*p*-dioxin (TCDD) in the diet. *Toxicol. Appl. Pharmacol. 50,* 241-252.

Ott, M. G., Holder, B. B., and Olson, R. D. (1980). A mortality analysis of employees engaged in the manufacture of 2,4,5-trichlorophenoxy acetic acid. *JOM J. Occup. Med. 22,* 47-50.

Pazderova-Vejlupková, J., Němcova, M., Pícková, J., Jirásek, L., and Lukáš, E. (1981). The development and prognosis of chronic intoxication by tetrachlorodibenzo-*p*-dioxin in men. *Arch. Environ. Health 36,* 5-11.

Pocchiari, F., Silano, V., and Zampieri, A. (1979). Human health effects from accidental release of tetrachlorodibenzo-*p*-dioxins (TCDD) at Seveso, Italy. *Ann. N. Y. Acad. Sci. 320,* 311-320.

Reggiani, G. (1980). Acute human exposure to TCDD in Seveso, Italy. *J. Toxicol. Environ. Health 6,* 27-43.

Zack, J. A. and Suskind, R. R. (1980). The mortality experience of workers exposed to tetrachlorodibenzodioxin in a trichlorophenol process accident. *JOM J. Occup. Med. 22,* 11-14.

DISCUSSION

DR. BARNES: Would you comment on the question of soft tissue tumors in the study in Sweden?

DR. CORDLE: The problem with this study is that these people were exposed to so many things that it is difficult to pinpoint the causative agent, similar to the case of dioxin. There were too many variables to attempt to do this. All this work has, however, been published.

DR. FRAWLEY: In your public health advisory to the Great Lakes area, the concept of reducing the amount of intake of fish as a food was introduced with the recommendation that perhaps fish should only be consumed twice a week rather than once a day. This is a new concept in tolerances, because we are accustomed to having tolerances in so many parts per million, etc. I hope this is not repeated in other areas of the Food and Drug Act.

DR. CORDLE: It should be emphasized that this was only advisory in nature and should be so considered; it was not intended as a regulatory control measure.

DR. KOLBYE: To set the record straight, this was not a tolerance--it is not even an action level. We do not believe that fish in interstate commerce in the United States are going to be a problem. One may refer to people who live in a particular area, eating particular species of fish and consuming large amounts of it. This was advisory in nature, particularly for the State of Michigan, and is very similar to previous advisories suggesting steps be

ISBN 0-12-193160-9

taken to limit the intake of fish. This has been done previously for mercury and for polychlorinated biphenyls. The importance of epidemiology in studying the patterns of anticipated human exposure with regulating food consumption as compared to regulating a food additive should be emphasized. There is a fundamental difference between banning a food additive and banning a good with all the economic and agricultural impacts that follow such an action.

DR. van RAALTE: In the Swedish studies mentioned, it should be noted that these are typical examples of articles that should be read, not in summary, but critically in their original form. Eight articles in six journals were included, but are all essentially the same work. Basically these are two studies and both are similar. One discussed railroad workers who had handled 2,4,5-T and the other the weed killer (Emitol) along the right-of-way. It was discovered that those exposed to Emitol had some excess tumors, but those exposed to 2,4,5-T did not. A reinvestigation of the same groups of people a few years later, showed a sixfold increase in stomach tumors. In addition, clusters of tumors in soft tissues were found that had not been detected before, another reason to perform additional studies. The results indicated a sixfold increase in tumors, but the problem encountered was that they were all different kinds of tumors of a very unusual pattern. This suggested that the cause of these unusual patterns could be something else, although it is felt that this cannot be true. These herbicides have been universally studied; in a few counties in Sweden, where the exposure was not excessive, an enormous increase in tumors was found. Again, it cannot be proved but it is extremely unlikely to say the least. The subjects of Dr. Kearney's studies should be followed-up in this country.

CHAPTER 17

POSSIBLE CONSEQUENCES OF SHARING AN ENVIRONMENT WITH DIOXINS*

Donald G. Barnes

Senior Science Advisor
Office of Assistant Administrator
for Pesticides and Toxic Substances
Environmental Protection Agency
Washington, D. C.

I. INTRODUCTION

In our experiences with polychlorinated dibenzo-*p*-dioxins (PCDD's), scientists have done a competent job in collecting facts. They have established the extreme toxicity of some PCDD's, suggested a variety of sources of these materials in the environment, and raised the concerns of tens of thousands of citizens who, before 1970, had never heard of "dioxins." In the process, however, has been the surprising discovery suggesting that we have not yet found the outer bounds of the "dioxin problem" or discovered how all the pieces fit together.

Facts are now becoming available which are beginning to help us understand what has been experienced thus far, although

**This chapter reflects the views of the author and not any agency with which the author is affiliated.*

ACCIDENTAL EXPOSURE TO DIOXINS

ISBN 0-12-193160-9

there remain many unanswered questions--some of which are very fundamental. This chapter will attempt to review, from the perspective of the EPA, some of the facts concerning herbicides, industrial wastes, abandoned dumpsites, controlled combustion, and area contamination to better understand what has been going on and what should be done in the future.

The term "dioxin problem" is often used. However, in some cases, what is actually being referred to is TCDD or, more specifically, 2,3,7,8-TCDD. This leaves open the possibility of misinterpretation, and perhaps creates greater anxiety in the public.

II. THE MANY FACES OF "DIOXINS"

The EPA is charged with administering the Toxic Substances Control Act (TSCA) and the Federal Insecticide, Fungicide and Rodenticide Act (FIFRA). A major criterion for action in each of these Acts is the finding of "unreasonable risk." Such a decision involves more than simply determination of risk. Rather, risk must be balanced against the benefits involved. Consequently, it is conceivable that a risk may be "reasonable", in some instances, and "unreasonable" in others, depending upon the associated benefits.

A. HERBICIDES

In the early 1970's it was thought that the "dioxin problem" was synonymous with TCDD, which was associated with the use of 2,4,5-TCP/2,4,5-T/Silvex. Later it was determined that the only compound of the TCDD group that merited special attention from an environmental point of view was the 2,3,7,8-isomer. This subject has engaged some of our most capable minds and has attracted the most attention in the media. The issue has been argued before an EPA Administrative Law Judge for more than a year. Approximately 50,000 pages of written testimony and oral cross-examination graphically speak to the amount of information which has been accumulated on the subject. This does not include the tremendous quantities of published literature that has also appeared on this subject.

Given the magnitude of this hearing and the fact that these administrative hearings are currently underway, it would be inappropriate to comment further on the issue of 2,3,7,8-TCDD in herbicides at this time, other than to note that the resolution of this issue in that forum is likely to leave many of the scientific questions about PCDD's, in general, unanswered.

B. INDUSTRIAL WASTES

The industrial wastes generated from the production of these herbicides contained concentrations of 2,3,7,8-TCDD up into the hundreds of parts per million. This material was incinerated, dumped, and, in some cases, stored. For example, in Missouri, the Syntex Corporation unwittingly purchased property containing a large volume of stored 2,3,7,8-TCDD. The Syntex Corporation has developed and operated a photolytic decomposition process for dramatically reducing the 2,3,7,8-TCDD content in the 2,4,5-T liquid production wastes produced by a previous owner. Successive treatment of the material has reduced the concentration many fold. The success of this effort was based on the fact that the wastes were in liquid form with a reasonably well-known composition. However, at another industrial site in Missouri, 2,3,7,8-TCDD-containing wastes of unknown and variable composition have been stored in hundreds of drums. The contents range from liquids to viscous semisolids. The challenge remaining is to either develop a treatment method that will dramatically reduce the dioxin concentration or to devise a disposal method which will safely segregate the wastes from the rest of the environment, thereby assuring that it will not become an abandoned dumpsite.

C. ABANDONED DUMPSITES

In the past chemical waste problems were sometimes handled by burying them in dumpsites with varying degrees of sophistication. The debate surrounding the passage of the Superfund legislation noted that there were many such sites in need of remedial action. In addition, there are an unknown number of other sites which have either been lost from a continuing tracking system or were clandestine from their very beginnings.

An investigation of a series of sites associated with Love Canal in New York, has revealed the presence of 2,3,7,8-TCDD in leachates from abandoned chemical wastes. The situation at this hazardous waste site was the subject of much litigation. One of the important issues raised at Love Canal involved the relative risks posed by (1) excavating 2,3,7,8-TCDD-contaminated earth versus (2) leaving the toxic-bearing material in place and dealing with any migration of the 2,3,7,8-TCDD-containing leachate. This is a decision which must be confronted at every 2,3,7,8-TCDD-containing landfill; no one option is likely to be optimal for all sites.

A similar situation was encountered in Missouri where, a number of years ago, drums of industrial wastes were buried on a farm several miles from any inhabited structures. In this case a remedial action was proposed by the Syntex Corporation, the present owners of the facility which first generated the wastes. Working with various units of the EPA, the company has been granted approval to excavate the site and recover the wastes for further treatment. A provision also exists for a long-term experiment to be conducted on the potential use of microbiological organisms in the decontamination of 2,3,7,8-TCDD-containing soils.

It is appropriate at this point to mention that the EPA has been actively seeking ways of "marshalling the microbe" to help deal with this problem. While such efforts in various parts of the world have been very successful, EPA-funded efforts have resulted in the creation of a microbe capable of existing on 2,4,5,-T as a carbon source. The goal is to discover a microbe that will utilize 2,3,7,8-TCDD as a significant part of its carbon source.

D. CONTROLLED COMBUSTION

In the fall of 1980, the EPA conducted careful stack sampling during trial burns at two incinerators in the Midwest, which were applying for permits to commercially burn PCB's. Earlier tests raised some questions about tetrachlordibenzofuran (TCDF). Analysis of the emissions revealed that 2,3,7,8,-TCDF was emitted along with a variety of other TCDF's and TCDD's. These materials were emitted under a variety of conditions: combustion of routine wastes, combustion of routine wastes plus PCB's, and combustion of PCB's plus clean fuel. As anticipated, the greatest emissions, however, occurred during the combustion of routine wastes plus PCB's. With this information in hand, the Regional EPA office made a series of "worst-case assumptions" which tended to maximize potential toxicity (e.g., treating all TCDD's and TCDF's as being as potent as 2,3,7,8-TCDD) and potential exposure (e.g., assuming that all the TCDD in or on fly ash recovered during analysis is biologically available to the organism). Careful consideration of the potential for carcinogenic and adverse reproductive effects lead to the conclusion that the incineration of PCB's could be allowed under controlled conditions at these facilities.

A second example of controlled combustion and PCDD's can be found in the combustion of municipal wastes. Analyses of 1979 emission samples taken from a municipal waste combustor revealed the possible presence of small amounts of TCDD's. In order to validate the conditions, a resampling and analysis was scheduled. Before this could be accomplished, however, data from Europe became available which reenforced the initial findings, i.e., particulate matter associated with combustion of municipal wastes can be contaminated with PCDD's and PCDF's. Analysis of the reported data led the Administrator to conclude that the situation did not appear to call for precipitous action. The Agency has now completed the sampling and analysis at three additional sites in the United States and plans to continue sampling and analyzing for the presence of PCDD's and PCDF's at selected municipal waste combustors during 1982 in order to better characterize the situation.

Mention should be made of the fact that the EPA has not detected TCDD in the fly ash from nine coat-fired power plants sampled. Additional plants of this type will be examined in 1982.

E. INADVERTENT COMBUSTION

Within the past year we have been sensitized to the possibility of the formation of PCDD's and PCDF's during the combustion of various materials. An electrical fire in the basement of a large office building in Binghamton, New York, earlier this year resulted in the contamination of much of the building with a thin oily soot. Analysis of this material revealed the presence of PCDD's and PCDF's in the parts per million range. State and local authorities are still struggling with questions of clean-up operations and public safety.

Since that time, analogous but less dramatic (in terms of location and extent), incidences have occurred elsewhere. Concern has been expressed by both firefighters and building managers about the likelihood of such occurrences and the appropriate actions to take. The states and local communities are dealing with a complicated problem, possibly ameliorated by the fact that these toxic pollutants are associated, in these cases, with particulates.

F. WIDE AREA INVOLVEMENTS

Within the past year a number of investigators both in the United States and Canada have reported finding 2,3,7,8-TCDD in fish and the eggs of herring gulls. The United States Food and Drug Administration (FDA) and the Canadian Health and Welfare have reviewed over a hundred analyses and have established levels of concern in the tens of parts per trillion range. The State of New York has independently conducted a review of these data and established a somewhat lower level.

The question is now being asked: Where did this material come from and what can be done about it? The Joint Commission on the Great Lakes is investigating these questions along with the allegation that this problem was created in the United States.

III. GAPS

It should be noted that only recently have these situations become known, although some of them have existed for some time. Some of this awareness is directly tied to the recent advances which have been made in the area of analytical chemistry and some of it is related to the fact that even since the technology has been available, no one thought to look at or ask questions about some of the places where we are now finding PCDD's. This awareness is also no doubt associated with the attention given "the dioxin issue" by the press and the depth of feeling that this issue can engender.

As we move forward in this area there are a number of important gaps that must be filled if the problems that remain are to be solved. At the same time the situation must be controlled, rather than vice versa.

A. BACKGROUND LEVELS OF PCDD'S AND PCDF'S

The question has been raised as to the likely existence of background levels of PCDD's and PCDF's in the environment. If such levels exist, they need to be analyzed and quantified. Such information will bear on answers to the following questions: How much of these materials have we been living with? How clean is "clean"? How persistent are these substances?

B. EFFECTS OF PCDD'S AND PCDF'S, OTHER THAN 2,3,7,8-TCDD

As other PCDD's and PCDF's are discovered in the environment, questions are raised about the toxicological properties of these materials relative to 2,3,7,8-TCDD. Considering the time and effort that has been expended in learning what is now known about 2,3,7,8-TCDD, it is clear that a similar assault on all of the other PCDD's and PCDF's would require a tremendous effort throughout this decade and much of the next. The situation is compounded by the lack of adequate samples of these compounds. Some steps are being taken by the FDA to alleviate this particular problem by synthesizing milligram quantities of all 22 TCDD isomers. This leaves open the question of most of the other PCDD's and PCDF's, however. Some other approach is obviously required.

C. MECHANISM OF ACTION

The mechanism of action of the acute lethal effects of 2,3,7,8-TCDD continues to elude our best minds. If this were known, the acute toxicity problems surrounding the other PCDD's and PCDF's could be resolved with relative ease. The question remaining would then be what mechanism of action would lead to other toxic end points.

While the task is formidable, much information can be gained from studying the great variations which have been observed in toxic responses between different species. There is the possibility that carefully conceived, conducted, and interpreted experiments will be able to short-circuit much of the work that lies ahead.

D. BIOAVAILABILITY QUESTIONS

In many instances in which PCDD's are encountered in the environment, there is an association with particulate matter. The adsorption/chemisorption of this material is very strong and will require considerable effort in the laboratory to remove them. Therefore, there is a need to determine the effect of this strong interaction on the toxicological properties of the PCDD's, including the potential for release to the organism under various conditions.

E. EFFECT ON HUMANS

Much attention has been focused on the potential adverse effects of PCDD's and PCDF's on humans. While mechanism of action studies will provide insights, additional information can be gleaned from epidemiological studies. A number of studies in the past 2 years have begun to focus the debate. The Dioxin Registry compiled by NIOSH seems to hold the promise of generating additional information on this particular portion of the problem.

IV. CONCLUSION

This brief chapter only partially summarizes the way "the dioxin problem" has grown and become more complex in such a short time. The present situation certainly is disconcertingly open-ended. Thus far we have been essentially lucky that the extreme toxicity demonstrated in animals may not have been manifested in humans. Although this has been fortuitous a fundamental understanding of what is occurring is needed. Thus, the problem is challenging and complex. A definitive solution will require our best efforts and with a cooperative attack on the problem, more progress can be made. This goal will be to enjoy benefits, without unreasonable risks.

DISCUSSION

DR. BARNES: There are six areas that illustrate what is known of the problem and what the needs will be in the future: (1) There is an increased awareness of the problems of dioxins. (2) The present analytical capability will make available additional information on which to take action. (3) Newspapers and related media will be able to properly inform the public as to the true nature of the hazards. (4) More data will be needed on background levels for all dioxins, and a general policy should be formulated regarding all the 22 isomers of dioxin since the effort devoted to the 2,3,7,8-tetrachlorodibenzo isomer cannot be duplicated for all isomers. In other words, we need to know more about the mechanisms of action of the dioxins. (5) The avenue of entrance to the body must be explored in more detail, e.g., air, water, and food. Finally, (6) we must discover if man has special resistance to these compounds as compared with animals. To achieve these goals it is imperative that there be a cooperative effort of the scientific and regulatory communities.

ISBN 0-12-193160-9

OPENING REMARKS

DR. COULSTON: The issue of a strategy for environmental reclamation and community protection can be addressed in great detail.

DR. BRUZZI: At the time of the accident in Seveso, over 100 children 8-14 years old, were exposed to the toxic chemicals. Chloracne was produced as a result, weeks and months later. This dermatologic effect was found on the back and was not actually related to unprotected or covered skin areas. Although this dermatologic finding was originally associated with TCDD, it is now recognized as caused by other chemicals. There is now a group of acneogenic chlorinated substances causing different degrees of dermatologic effects. TCDD seems to be more active in this respect than others in this category. The mode of action is unknown, but new data is now available on the subject. Some of these chemicals cause an acute, red dermatitis that is different from chloracne. Chloracne can even follow ingestion of these compounds. In Seveso, a few chloracne cases were diagnosed in 1977. Fewer were diagnosed in 1978, but none later. There were no sex differences. At the present time, two females and one male still have the chloracne. To summarize, the chloracne could only have been caused by exposure to the air and to the clouds, and no other acneogenic substances but TCDD.

DR. COULSTON: The chloracne issue as understood is very clear --it could be due only to dioxins. However, could it not be due in part to other substances that might have been emitted into the air from the factories?

ISBN 0-12-193160-9

DR. POCCHIARI: There was no other acneogenic substance. Therefore, there is no doubt but that there is a causal relationship between the data and the incidence of chloracne. At this point the dosage that produces chloracne is known. That this occurs has been shown by Professor Schulz by applying this substance on his skin.

DR. BRAUN: In 1966-1967, Professor Klingman similarly applied 1-7500 μg of this substance on himself. The higher doses produced severe chloracne, but the 1 μg doses did not. The measurements showed no change in liver function even at the high doses.

DR. REGGIANI: In this connection it is noteworthy that following the Monsanto 1949 Nitro, West Virginia accident where about 250 people were exposed to probably very high doses during the production of 2,4,5-T, about 50% of the people exposed now have chloracne. In similar experiences in other countries about 50% of the exposed people also have chloracne. It is not known why the chloracne continues, but presumably the substance is retained for years in the subcutaneous tissues. Recent findings in monkeys may support the concept of long-term body residues because it appears that in these animals the half-life is about 1 year.

DR. SHEPARD: This issue of long-term retention of TCDD in the body as related to dermatitis and other effects is difficult to resolve. It may be similar to other sensitizing agents like poison ivy that have a delayed effect for onset of dermatitis and a prolonged persistence. This kind of dermatitis can also be caused by oral ingestion. It may not be necessary for the original chemical substance to be retained as such and the half-life, even if it is 1 year, would suggest that most of the original chemical would be gone after six or seven half-lives. However, the original tissue cellular changes may still persist even though the initiating chemical is no longer there.

DR. HOBSON: It may be fruitless to speculate on these points until more facts are known. Because there are so many sources of TCDD, it is really impossible to prove cause as related to any specific source.

CHAPTER 18

BIRTH DEFECTS IN THE TCDD POLLUTED AREA OF SEVESO: RESULTS OF A FOUR-YEAR FOLLOW-UP

Paolo Bruzzi

Instituto di Oncologia
Università degli Studi
Genova, Italy

I. INTRODUCTION

The procedures of Seveso Defect Register (SBDR) have been previously described in detail. The Registry was established during 1978. Data for 1977 was collected retrospectively. Table I provides some of the data collected on the populations of the 11 municipalities and the four zones (A,B,R,"Out") which have been accumulated using data derived from this Register.

II. RESULTS

From the data collected during the period from 1977 to 1980, the average number of births has increased from 86 to 94% with little variation among towns. The number of newborns with birth defects has also been determined. Diagnosis of these defects is now more accurate, thus making the data more reliable (Table II).

In the analysis of birth defects the SBDR data have been compared with those from other registries for two purposes:

ISBN 0-12-193160-9

TABLE I. Areas and Inhabitants (Residency on May 31, 1981) by Municipality and Pollution Zone

Municipality	*Area (ha)*	*Population*
Barlassina	287	5,592
Bovisio	493	11,129
Cesano M.	1,146	32,665
Desio	1,479	33,659
Lentate	1,399	13,162
Meda	834	20,260
Muggio'	547	18,457
Nova M.	581	19,210
Seregno	1,301	38,005
Seveso	734	17,620
Varedo	484	12,242
Total	9,285	222,001

Zone	*Area (ha)*	*Population*
A	108	670
B	270	4,855
R	1,430	32,481
Out	7,474	183,995
Total	9,285	222,001

(1) to test their overall reliability and (2) to identify those, presumably few, deviations that may be related to TCDD exposure.

The rates for selected malformations derived from the SBDR data fall within internationally established ranges, thus indicating the reliability of the SBDR. For four types of malformations, however, the rates fall outside this range. Two such deviations, one case each of rectal-anal atresia and cleft palate are attributable to the small sample size of the SBDR. (The excess number of cases observed for hypospadias will be discussed later.) The greater number of cases of Down's Syndrome approaches the rate observed in most Latin countries.

TABLE II. Birth Defects Registry during the Period 1977-1980 by Municipality and Pollution Zone

Municipality and zone	*1977*		*1978*		*1979*		*1980*	
	Births no.	*Known %*	*Births no.*	*Known %*	*Births no.*	*Known %*	*Births no.*	*Known %*
Barlassina	63	87.3	55	76.4	61	93.4	44	88.6
Bovision	132	97.0	139	95.0	128	96.9	101	96.0
Cesano, M.	362	88.4	419	91.9	366	96.4	347	91.1
Desio	411	86.4	432	93.5	370	94.6	392	92.9
Lentate	147	78.2	173	90.2	134	91.8	125	80.8
Meda	279	88.2	286	87.8	228	91.2	248	89.1
Muggio'	255	85.1	245	89.8	212	91.5	198	87.9
Nova M.	296	86.5	314	90.1	238	92.0	275	89.8
Seregno	457	82.3	439	86.8	443	95.5	454	90.3
Seveso	188	88.3	220	90.4	205	96.6	187	90.9
Varedo	159	89.3	154	92.9	122	95.9	135	92.6
Total	2,749	86.4	2,876	90.3	2,507	94.4	2,506	90.3
Zone A	6	66.7	3	66.7	11	90.9	1	100.0
Zone B	67	85.1	85	89.4	86	97.7	67	95.5
Zone R	408	91.4	438	92.0	393	97.2	430	92.8
Out	2,268	85.6	2,350	90.0	2,017	93.7	2,008	89.6

TABLE III. Selected Congenital Anomalies Reported in SBDR and in Other Lombardy Registries[a]

	Lombardy 48,828 Oct. 1977-June 1980		*Seveso 10,638 Jan. 1977-Dec. 1980*			
Anomaly	*Cases*	*Rate (X10000)*	*Cases*	*Rate (X10000)*	*Excess or defect*	*P value*
Anencephaly	12	2.46	4	3.76	+	0.46
Spina bifida	23	4.71	10	9.40	+	0.10
Hydrocephalus	16	3.28	5	4.70	+	0.67
Neural tube defects	51	10.44	19	17.86	+	0.063
Cleft palate	21	4.30	1	0.94	-	0.17
Total cleft lip	32	6.55	7	6.58	=	~1
Esoph. atresia	25	5.12	1	0.94	-	0.11
Rectal-anal atresia	27	5.53	1	0.94	-	0.084
Hypospadias	11	2.25	30	28.20	+	<0.0001
Pes equinovarus	62	12.70	13	12.22	=	0.90
Polydactily	24	4.91	11	10.34	+	0.061
Syndactily	18	3.86	3	2.82	-	0.67
Limb reduction def.	44	9.01	3	2.82	-	0.061
Downs Syndrome	72	14.75	18	16.92	+	0.70

[a] *Data reported for two groups of I.P.I.M.C. hospitals.*

TABLE IV. Distribution of Congenital Malformations at SBDR by Year

Year	Cases (children)	*All malformations* No.	*All malformations* Rate (x10000)	*Skin anomalies excluded* No.	*Skin anomalies excluded* Rate (x10000)	Proportion of skin anomalies (%)
1977	87	92	334.67	67	243.72	27
1978	143	146	507.65	98	340.75	33
1979	157	168	670.12	81	323.09	52
1980	124	127	506.78	59	235.43	54
Total 1978-1980	511	533	501.03	305	286.71	43

In fact, it falls within the Italian range and close to the Lombardy rate, as will be shown later.

SBDR rates were compared to the rates obtained from various Italian hospitals reporting to the Multicentric Survey of Birth Defect. Unlike the comparison with international ranges, the definition and classification criteria for each selected malformation were homogenous within the survey, thus, the ranges were narrower. SBDR rates for esophageal and rectal-anal atresia (one case each) and for limb reduction deformities (three cases) were below the reference ranges, presumably due to our small sample. The slight deviation observed for cleft lip and palate was possibly due to a difference in classification criteria. The excess number of cases observed for neural tube defects, however, could be dismissed by either one of the above reasons.

Table III shows the number of cases and the rates reported by a group of hospitals in the Lombardy region (where our Registry is located) and by SBDR. The *P* value for each difference is given. The previously noted deficit from the SBDR data was confirmed for limb reduction deformities (3 cases observed against 9 expected), and for cleft palate, esophageal, and rectal-anal atresia (1 case observed and 5 expected for each). The excess number of cases of neural tube defects appears mostly due to spina bifida (10 cases observed and 4 expected), while the highly significant number of cases of hypospadia (also in excess of the international range) was only partially accounted for by differences in diagnostic criteria. For polydactily, however, 11 cases were observed against the 5 expected.

While external comparisons are affected by the various operations of each Registry, internal comparisons are not. When our data are stratified by time or zone of occurrence, the data obtained are most severely affected by the small size of the sample and by the uncertainties inherent in the assessment of TCDD exposure.

The number of malformations observed at SBDR by year is shown in the Table IV. As previously noted, the reliability of SBDR data for 1977 is not as good as in the 3 following years (92 cases as compared to 127-168). The extent of underreporting in 1977, however, probably does not affect all types of malformations. Skin abnormalities (largely hemangiomas) are less than 30% of the total in 1977 and more than 50% in 1979-1980. If such abnormalities are excluded, overall malformation rates show a clear declining trend from 1978 through 1980.

The annual rates observed at SBDR for two selected groups of malformations show one group exhibiting an increasing trend, while another showing a declining trend. To more definitely identify a trend, however, if such trend exists, would require a much larger population base. It would also require that oc-

TABLE V. SBDR Malformation Data According to Zone of Residence of Mother, 1977-1980

Malformation	*Live and still births*				
	A+B zone (326)		*R zone (1669)*		*Out zone (8643)*
	No.	*(RR)*	*No.*	*(RR)*	*No.*
Nervous system anomalies	1	(1.3)	2	(.5)	21
Anomalies of the eye	-		2	(.9)	11
Anomalies of the ear	-		2	(.4)	24
Branchial cysts	-		-		2
Heart anomalies	4	(3.1)	6	(.9)	34
Anomalies of circulatory system	-		2	(1.5)	7
Anomalies of respiratory system	-		-		4
Cleft lip and cleft palate	-		-		8
Anomalies of digestive system	-		4	(2.6)	8
Anomalies of genital organs	3	(2.9)	5	(.9)	27
Anomalies of urinary system	-		-		5
Anomalies of musculoskeletal system	1	(.4)	12	(.9)	68
Anomalies of closure of abdominal cavity	-		1	(1.3)	4
Anomalies of integument	10	(1.5)	43	(1.2)	177
Chromosomal anomalies	1	(1.5)	3	(.9)	18
Other congenital anomalies	2	(6.6)	3	(1.9)	8
Total malformations	22	(1.4)	85	(1.0)	426
"Hemangiomas excluded"	13	(1.2)	50	(.9)	277

TABLE VI. Multiple Malformations From 1977 Through 1980 by Pollution Zone (Excluding Known Syndromes and Associations with Hemangiomas

Zone or area	*1977*	*1978*	*1979*	*1980*	*Total*			
					Observed	*Expected*[a]	*R.R.*	*°/ooo*
A + B	0	1	1	0	2	.56	3.58	61.35
R	0	3	2	1	6	2.86	2.10	35.95
Out	3	4	5	3	15	14.81	1.01	17.35
Italy (169,142 births from October 1977 through June 1980)					290	//	//	17.14

[a] *Expected values calculated on the basis of Italian rate.*

currences, such as the accident at Seveso, be more frequent. It may be presumed that skin and ear abnormalities, mostly slight and mild, were underreported at the time when the SBDR operation began. Such a presumption is not sustainable, however, for multiple malformations (to be discussed later). A declining pattern is visible for cardiovascular, and possibly for musculo-skeletal and genital anomalies.

Table V presents an analysis of SBDR data according to the zone of residence of the mother for each group of malformations. Such an indicator for potential TCDD exposure as compared to birth defects, however, is unsatisfactory for two reasons. First, zoning criteria have been severely criticized since the epidemic of chloracne (134 cases in Zones A,B, and R, and 53 out) and the observed animal deaths spread well beyond the boundaries of the established zones. In addition, this data was obtained on the basis of soil assays only. An attempt to draw a more reliable exposure map on the basis of *all* available criteria is currently underway. Second, due to various mobility, behavioral, and social patterns, the actual human exposure was inherently much more difficult to assess than animal exposure and less related (than the latter) to environmental pollution maps.

With these limitations in mind, the analysis by zone shows some correlation between potential TCDD exposure and occurrence of some malformations, such as those affecting heart and vessels, the genital system, the integument, and the "other", which include identifiable syndromes. A slight excess of risk was also observed for all malformations, at least in the most polluted Zone B.

Table VI shows children with multiple malformations by year and by zone of occurrence as compared with Italian data. Even though the small number of cases involved precludes a meaningful statistical testing, the correlation observed between potential TCDD exposure and relative risk is suggestive.

III. DISCUSSION

Table VII summarizes the observations presented here concerning those anomalies which have shown some remarkable features. Concerning the anomalies with less-than-expected incidence (shown at the bottom), underreporting seems an unlikely explanation, due to their rather obvious nature. The small number of cases observed by SBDR data precludes any internal comparison and requires further investigation.

A possible association with TCDD exposure is not supported by the overall evidence concerning polydactily and Down's Syndrome. It is, however, suggested by the available evidence

TABLE VII. Summary Evaluation of Selected Anomalies

Anomaly	*External comparisons*	*Internal comparisons*	
		by zone	*by year*
Heart+vasc.	?	++	Decrease
Hypospadia	+	+	Decrease (?)
Neural tube defect	+	=	No trend
Multiple	?	++	No trend
Down's syndrome	(+)	=	No trend
Hemangioma	?	+	Increase
Polydactily	(+)	=	No trend
Esoph./anal rectal atresia	-		
Limb red. def.	-	?	?
Cleft palate	-		

concerning hemangiomas and, perhaps, neural tube defects. It is more definitely indicated for hypospadia, because of a strong consistent excess over other registries and a weak correlation with potential TCDD exposure, and for both cardiovascular and multiple defects, because of a reasonably good correlation and a possible time trend. Further studies are needed to verify such a hypothesis.

CONCLUDING DISCUSSION

Chairman: *F. Coulston*

DR. COULSTON: It is agreed that for the statistical data presented, the records are not as numerous or as accurate as desired. However, the data do show that there is no change in the abortion rate related to TCDD exposure. There must be some psychological bias that influences the reporting of abortions. It appears that people begin reporting many ill effects after realizing that they have been exposed to a toxic substance like TCDD.

PROFESSOR TUCHMANN-DUPLESSIS: Most experienced investigators do not attempt to equate data on miscarriages from one region to another. For this reason, it is difficult to accurately evaluate the findings in Seveso, because no such epidemiological studies were undertaken before the accident in this area. In any event, because of the small number of cases any inference of cause and effect relationship is difficult. However, for correct epidemiological records, miscarriage is the proper term. The world-wide data on miscarriages are difficult to obtain and are likely incorrect, because people do not want to admit this.

DR. COULSTON: These data need to be very carefully examined, because some people may say there is a dose-response relationship.

ISBN 0-12-193160-9

PROFESSOR TUCHMANN-DUPLESSIS: There should be some discussion on the details of toxicity to the early embryo. An embryo can be killed for many reasons and embryotoxicity is a well-recognized effect caused by many situations and many substances. The question of embryo fatality should remain open and the question of congenital malformations be discussed instead.

DR. FRAWLEY: It can be said that because of the heightened concern in the minds of these people in the Seveso areas, that we have paid more attention to the abortion rate. However, when the figures were derived, it was found that they actually fell within the normal range--accurate calculations placed the figures well within the world average. Is this correct?

DR. COULSTON: Yes, this is essentially correct.

PROFESSOR TUCHMANN-DUPLESSIS: The Seveso data were collected 1 year later, but should have actually been started within the first year. The first trimester is the most critical period for fetal injury by chemicals or any other adverse effect. Therefore, it is essential that prior records be available. Accurate data are necessary for the first 3-6 months following exposure, and full recognition must be given to all the complicating factors that influence reporting of miscarriages and birth defects. The legal complications alone are enough to confuse the data collected. I believe that TCDD exposure does not cause malformations, but the data are too few to be able to draw conclusions about the potential future teratogenic action of TCDD exposure. There may be some little recognized malformations, but this is hard to prove.

One point that suggests that TCDD is not a strong teratogenic agent is the finding of so many different types of teratogenicity in the fetuses studies. Had TCDD been a potent teratogen it would have produced many malformations of the same kind in even the small number examined. Thus, we can be sure TCDD is not a strong teratogen as compared with other known agents. There is one comforting conclusion: this study is now going on and will be followed-up in the children now and in the years to come. Future findings will no doubt answer many of the questions raised today.

DR. COULSTON: Are there any data on conception of women over time? Have children been born to mothers who were exposed several years ago?

DR. BRUZZI: This is a part of the study now going on, and full information is not yet available.

PROFESSOR TUCHMANN-DUPLESSIS: How many children have been born after the accident having been exposed *in utero* during the first 3 months of pregnancy?

DR. BRUZZI: For both Zones A plus B there were about 350. Of these, there were 22 who were considered deformed.

PROFESSOR TUCHMANN-DUPLESSIS: This statistical evaluation therefore needs to take into consideration minor malformations, because this would be necessary to reach this 6% level. It is important to really consider only true, major malformations. The point is the heterogeniety of the malformations. It seems that there is no significant increase in birth defects related to the exposure from this accident.

MR. HUTT: Dioxins should also be discussed from a regulatory point of view. They do not present regulatory issues at this time, but as can be seen do give rise to many questions of a scientific nature. If it can be agreed on a scientific basis that there is a public health concern, then regulatory decisions are clear.

Dioxin is not unique and in many ways resembles the problems presented by the aflatoxins and nitrosamines. What is different between a natural contaminant and an artificial food additive? Both are equally difficult to get rid of, but for different reasons. Regulatory control is equally difficult and societal choice may become the overriding factor in the final decision. As our capabilities for toxicity testing and analysis becomes more refined the regulatory problems are going to increase. Dioxin is only one of the compounds which will have to be dealt with.

A special problem for regulation is the widespread lack of agreement on a single interpretation of the same data by scientists which is especially evident in any regulatory agency. It would be difficult to find a single case where there was total agreement among the scientific community. One example is the difference between the United States and Canada with respect to their regulations on Red No. 2 and Red No. 40. In Canada, Red No. 2 is regarded as safe and Red No. 40 is prohibited. In the United States this situation is reverse--Red No. 2 is prohibited and Red No. 40 is regarded as safe. It is very difficult to be consistent in a regulatory sense when there are such disagreements.

Disagreement also exist as to the appropriate governmental response when a problem such as dioxins and similar public issues arise. During the past 5 years the Congress of the United States has enacted more statues to override action by the FDA then it has enacted statutes to give it authority to take action. The FDA has been prevented, for example, from acting on vitamins and minerals, shellfish, saccharin and, if it had come to a final debate, nitrates in meat. Public concern and pressures often influence legislation that defeats the goals of the regulatory agency. Such statutes are readily enacted, when there is general agreement by the scientific community, the public, and Congress on specific health and safety issues. There are numerous regulations that include such substances as the dioxins. However, what is required is some better agreement among scientists in the guidance needed toward making these regulatory decisions.

DR. *BARNES:* By and large there is good agreement on a number of facets of the scientific debates on the dioxins.

DR. *KOLBYE:* Within the FDA for specific regulatory purposes 1 ng/kilogram body weight was considered a safe level for TCDD for fish only. However, degrees of safety differ depending on the nature and type of exposure. Thus, the EPA does not consider this level as safe. Instead, their criteria is risk and benefits, which may change in a specific case. Numerous factors influence such decisions on an individual case-by-case basis. Dioxins are considered contaminants and not used by man as an intentional additive to food or the environment. Thus, the control required is fundamentally different. In addition, TCDD is not a direct carcinogen and therefore, the conservative linear extrapolation model to calculate a safe dose should not be used. Dioxins may be considered promoters, but this question has not been completely answered.

DR. *COULSTON:* In the history of the toxicological experience with DDT, for example, in spite of the original restrictions that had been placed on its use, the evidence thus far accumulated is that it is not harmful to man, even after 40 years of use. One could argue that in the case of TCDD these results may be similar. A regulatory decision cannot and should not be made until there is agreement whether or not this is a harmful substance. In the meantime governments may, for whatever reasons, decide to prohibit or permit the use of any substance. The experience in the United States and Canada regarding the use of the dyes Red No. 2 and No. 40 is a case in point. Numerous similar differences exist between countries with respect to regulation of suspected toxic substances. It has been proposed that international agreement be required on the hazards of many substances. Organizations such as the International Academy of Environmental Safety and the others are attempting to work out some rational decisions. However, the international difficulties are so great, the substances involved so numerous, and the economic issues so enormous, that progress is slow. On the other hand, there is every reason to believe that in time, probably years, many of these ecotoxicological and environmental safety issues will be resolved. The history of compromise between nations certainly suggests that accord can be reached. These toxicological questions are so new and so technical that time is required for people and nations to adjust their thinking.

There should be a final discussion of other questions that have arisen. What is a no-effect level for dioxins? What is the regulatory or even scientific difference between promoters of carcinogenesis and initiators? How does one evaluate a reversible effect in animal studies? Is there a "safe" level? What is the position of FDA and EPA with respect to these questions?

DR. KOLBYE: No substantial hazard to animals is seen with dioxin at a feeding level of 0.001 μg per kilogram of body weight fed for a lifetime. At the same time, it is recognized that all the numerous facets of animal sensitivity, etc., that can be asked, will modify any conclusive figure. Aside from the data presently available, it appears that there is no substantial hazard in these animals at this 0.001 level. Whether or not one is willing to use this information to determine safety or not depends on many points and also the margin of safety required and the life situation under consideration. The type of human exposure will be extremely important in any final determination of a safety level for man.

DR. MATSUMURA: With respect to the difference between promoters and initiators of cancer, why do not the regulatory agencies make a distinction with respect to the level of risk from these different kinds of substances?

DR. KOLBYE: One reason is that even the scientific community does not agree about the nature of, or action of, or mechanisms involved in, the effects of these substances. Although it is known that there are major differences in the potency of many carcinogens, the reasons are not known. In addition, there is the ever-present confounding fact that the cause of neoplasia is unknown.

One of the difficulties in resolving these issues is that many of the more than 4000 references in the scientific literature dealing with very important research findings have been almost totally disregarded or ignored by most people involved in cancer research or the regulatory decisions made in the United States. An inter-agency working group will be formed to address this matter, and to collect and make available all this evidence on chemical carcinogenesis.

At this point we must differentiate between potentiators and initiators. An *initiator* is a substance that acts through some biological observable phenomenon to produce an increase in cancer risk. To the extent that some of these observed effects are reversible through biological changes, the dogma in this country has been that since cancer is not reversible, the causation of cancer is not reversible. There are now very substantial and sufficient reasons to question this dogma. With regard to *promoters*, there is a stage that, until it develops, events are reversible. The only way to explore this difficult question is to use intermittent dosing experiments with serial sacrifice. The progression of events may be illustrated in animals by a practical experimental trial of this character. In other words, there is a need to test in practice this reversible concept.

DR. COULSTON: The last feature of this discussion should be on what to do when an accident occurs and what reclamation possibilities there are when dealing with dioxin contamination.

DR. KEARNEY: Recently, the decontamination plan for the Seveso area was reviewed. At the time, a plan for immediate action was needed. Some actions proved to be good and extremely useful, while others such as incineration of contaminated articles, proved to be undesirable. It was necessary that a fence be erected around the whole contaminated area. This was achieved with a high, woven-wire fence with a canvas lining. Because people were entering and exiting the Zone it was thought that a fence was needed to block access. It was even important to prevent animals from coming in and out of this area, since rabbits caught several miles away, for example, contained high residues, and were being used as food. The canvas on the fence was useful in preventing dust from escaping the region. A safe disposal pit was constructed and this proved to be successful. A pass was required to enter Zone B. Although people wished to return to their homes for various personal reasons, access was restricted.

Another priority was to establish guidelines for decontamination based on residue data. In other words, how could an area be decontaminated and, further, what guidelines would be used to determine when that had been accomplished. The guidelines established for safe and unsafe areas were based on analytically determined values in plants, soil, and on building surfaces. These proved to be very satisfactory and workable. Residues of 0.1 to 1 ppm were considered unacceptable. Residues in the range of 10 to 100 ppb, were considered unsafe and decontamination was thus needed. Residues of less than 10 ppb were acceptable except for pregnant women and children. The ultimate goal was to reduce levels to about 10 ppt.

Decontamination procedures included destroying all buildings in Zone A for both scientific and socioeconomic reasons. Trees were cut down to remove contaminated leaves and vegetation. All livestock was removed. Flooding of the soil was not successful as an anaerobic measure to accelerate decomposition of the TCDD. Because rabbits were extremely sensitive to TCDD they were placed in small fenced-off areas and served as good indicators of when an area might be considered safe for human beings. All in all, in the face of extreme difficulties and pressure, the plan was successful.

DR. *BARNES:* May I ask for some specific example? Building decontamination poses special problems. Can surfaces be coated with a stucco-like material or with activated carbon in a slurry on the surface on top of that?

DR. *POCCHIARI:* Building decontamination is the most difficult kind of reclamation problem. Most surface coatings give only a low measure of temporary protection, because eventually the TCDD emerges through even the heaviest coatings. Outright destruction of a building is not really very satisfactory, because one must then dispose of the debris.

DR. *COULSTON:* What would EPA do if a Seveso-like episode happened in the United States?

DR. *BARNES:* A national contingency plan has been initiated in the United States to meet major emergencies from chemical spills and similar accidents. For example, in a recent accident in San Francisco gas mains were broken with the subsequent release of PCB's. As a result of activation of the national plan, within 3 days the situation was partially restored to normal. Unfortunately, the EPA or the national emergency plan does not include any way to decontaminate a building that has been contaminated with PCB's or TCDD. Actually, some methods of decontamination may only serve to drive the agent deeper into building surfaces from where it may emerge at a later time.

DR. *KIM:* New York State has a plan now to meet PCB accidents as a result of their past experiences. At the least there is an emergency stand-by plan of action, by which an immediate response team is activated. This does not necessarily mean the actions taken will be extraordinarily effective in terms of eliminating the contamination. However, it does illustrate an awareness of the need for prompt agency action. People will always suspect some injury at a later time and the postaccident legal aspects will go beyond the immediate protective measures taken.

DR. *SILANO:* The need for prompt action after an accident is well recognized. For example, it was 10 days after the Seveso accident before it was known that dioxins had been released into the environment. Although people were the primary concern there was an apparent lack of any well-organized plan of action. This experience has taught us an important lesson in that there is now an awareness of what needs to be done the next time there is a similar accident. Recognition of the value of technical people in assessing the extent and nature of the contamination is of utmost importance. Obviously, prevention of an accident of this type is best achieved by education, or to be sure that such episodes will not occur in the future. However, the actions to be taken after an accident will be based only on good planning.

DR. *COULSTON:* An overview of the discussions presented is obviously warranted at this point.

DR. *POCCHIARI:* Five years ago the accident which occurred in Seveso appeared to be a disaster. At the present time, it is not so much considered a toxicological as it is considered a public disaster, specifically in terms of people, buildings, etc. From a health point of view, it seems that the toxicological effects are not as great as was feared at first. The dioxins were extremely toxic to animals and it was feared the effects on human beings would be equally bad. With the exception of chloracne, the effects appear to be limited. Thus, 5 years later, the situation seems to be under control. We cannot discuss 25 years hence; this must be left for the future. From an ecological and financial point of view, the accident was a disaster. At least $100 million was spent not only by the government but by the people as well, not to mention the tremendous impact this incident had on the environment. The newspaper publicity helped to create a strong emotional public response. However, with due consideration, the actions that were taken, could be repeated should another accident occur in the future. There is one important lesson, however. In the past, actions taken were closely related to the pressures exerted by newspapers, than scientific data. This is something that must be changed. There must also be wide cooperation not only in the scientific community, but within world-wide regulatory agencies. Hopefully, from the data acquired here, other groups in other countries may reach similar rather than different conclusions.

For the future, it is important to have a plan. It is important to know who will be called to the accident first. There is a need for a general information center to call for help. The plans must be wide enough in scope so that all who are involved are properly represented. There must be general agreement about the decisions made. Every country in Europe needs a similar plan. There should be available experts who could advise other countries on how best to deal with strange accidents for which there is little experience.

There are certain basic, general plans that must be considered. Chemical plants are often the source of explosions that may call for immediate action. The same may be true of trains, ships, trucks, and pipelines. Contamination of food presents still another kind of toxic problem. Water supplies must be protected against contamination. All actions must be taken immediately. Nineteen days were required for completion of emergency actions after the Seveso accident. An emergency plan would have saved valuable time. In recent years there have been four accidents; it took far too long for a proper response for each one.

It should be suggested that an international group be established to formulate plans in dealing with emergencies. Such plans should include suggestions and ideas on how best to reclaim an area after an accident. There are many possibilities and many good ideas have been suggested.

DR. COULSTON: In conclusion, I wish to thank all of the speakers and discussants who have presented here for the first time a strategy for what can be done to correct a disaster such as occurred in the Seveso area, which can be adopted as a plan for all accidents of this kind. The meeting was told that the government of the United States of America has set up a national contingency plan to meet major accidents, resulting from chemical spillage. The major need is to work out a plan for decontamination of chemicals, whether they be PCBs or TCDDs along the lines which have evolved in New York State, as presented in this report. This meeting has contributed greatly to the understanding of the hazards and risks resulting from exposure to the dioxins. Dioxin is indeed a very dangerous chemical having no value to mankind and therefore must be controlled. However, it is in the environment and we must learn how to live with it. We must set human, plant, animal, and health standards at a level that we can live with, such as was done by the U. S. Food and Drug Administration in the case of fish. Dioxin cannot be entirely eliminated from our environment and, therefore, we must set standards for its presence and continue to evolve scientific and engineering methods to control it. The important thing is to develop scientific facts that enable us to set these standards to protect plant, man and animals from the harmful effects of dioxins. This strategy for environmental reclamation and community protection in terms of human health to accidental exposure to chemicals must become the immediate goal of all governments and international organizations.

INDEX

A

Agent Green, TCDD in, 235
Agent Orange, TCDD in, 229–230, 235
Agent Pink, TCDD in, 235
Agent Purple, TCDD in, 235
Agriculture, TCDD problems in, 233–240
Animals, TCDD levels in, at Seveso, 27–29, 30–31
Arkansas, 2,4,5-T exposure studies in, 237
Atmospheric particles, tetrachlorodibenzo-*p*-dioxin in, at Seveso, 16, 21–23

B

Bacillus megaterium, TCDD metabolism by, 127, 130, 131, 133
Binghamton (NY), PCB fire at, dioxins from, 191–195
Birth defects, from Seveso incident, 218–221, 271–280
Blood, halogenated doxin effect on, 57–59
Building reclamation, after PCB fires, 185-189

C

Cancer, incidence in Seveso area, 221
Carcinogenesis, mechanisms related to TCDD, 191–195, 286
Child development, halogenated dioxin effects on, 54–57
Chloracne
  incidence of
    at Nitro, W. Virginia, 270
    at Seveso, 225, 269
Chlorinated dioxins, formation of, 172
Clorobenzenes
  from incineration of polyethylene, 179
  PCDDs form from pyrolysis of, 177
Chlorophenolic pollutants, reactions of, 140–142
Combustion
  controlled, TCDD from, 262–263
  inadvertant, PCDDs from, 263
Cytochrome P-450 monoxgenase systems, in TCDD metabolism, 90–93

D

2,4-D
  exposure studies on, 237–239
    reproductive mortality, 239
    spontancous abortion, 239–240
Decontamination, after Seveso incident, 286–287
Dermatologic signs, of halogenated dioxin exposure, 52, 59, 217
Dioxins
  halogenated, *see* Halogenated dioxins
  tetrachlorodibenzo, *see* Tetrachlorodibenzo-*p*-dioxin
Dumpsites, abandoned, TCDD in, 261–262

E

Ecological chemistry, of dioxins and related compounds, 171–189
Environmental reclamation, examples of experiments on, 151
Epidemiological studies, on dioxins in food supply, 247–248
Exposure to halogenated dioxins, determination of, 41–43

F

Fertility, TCDD effects on, 207
Fish
  human exposure to TCDD in, 250–253
  TCDD in, 263–264
Food supply, regulation of dioxins in, 245–256

G

Great Lakes, TCDD in fish from, 252–255

H

H donors, in halogenated dioxin degradation, 155–156
Halogenated dioxins, *see also* Tetrachlorodibenzo-*p*-dioxin (TCDD)
  animal data on, 204–208, 248–250
  clinical symptomatology of exposure to, 49–54
  consequences of sharing environment with, 259–270
  effects on
    blood and tissues, 57–59, 60–61
    metabolism, 57
    pregnancy and child development, 54–57
  ecological chemistry of, 171–189
  epidemiological studies on, 247–248
  in food supply, regulation, 245–256
  health effects of, 39–67
  human exposure to, 250–252, 265
    from fish, 252–255
  mortality from, 218
  from PCB fires, 185–189
  photochemical degradation of, 149–158
  regulatory aspects of, 283–284
  risk assessment of low doses to, 43–44
  teratological aspects of, 201–214, 282
Health, dioxins effect on, 39–67
Hepatocytes, TCDD metabolism in, 88–90
Herbicides, dioxins in, 260
Herring gull eggs, TCDD in, 263–264
Humans
  exposure to TCDD in fish, 252–255
  halogenated dioxin effects on, 39–67

I

Industrial wastes, TCDD in, 261
Isopredioxin, formula of, 175

J

Japan, exposure to halogenated dioxins in, 45–64

L

LD_{50} values, for TCDD, 83
Liver cells, TCDD metabolism in, 88–90
Love Canal (Niagara Falls), human exposure to TCDD at, 45, 261

M

Metabolism
  halogenated dioxin effects on, 57
  of TCDD in mammals, 81–100
Michigan, human exposure to polybrominated biphenyls, 44
Milk, TCDD levels in at Seveso, 29–30
Microbial degradation of TCDD, 105–135
Microsomes, hepatic, TCDD metabolism in, 88–90
Missouri, abandoned dumpsites in, 261–262
Mortality from Seveso incident, 218–221

N

Neurological signs, of halogenated dioxin exposure, 53, 223
Nocardiopsis, TCDD metabolism by, 128, 129, 133

O

Oxidative control of chemical pollutants, 139–147

P

PCDDs
  background levels of, 264
  from controlled combustion, 262–263
  effects of, 264–265
Pentachloro isomers, dissipation time of, 59
Photochemical degradation
  of halogenated dioxins, 149–158, 173
  of TCDD, 180
Plant tissues, TCDD levels in, at Seveso, 23–27
Pollutants, ruthenium tetroxide in oxidative control of, 139–147
Polybrominated biphenyls (PBBs), exposure to, in Michigan, 44
Polychlorinated byphenyls (PCBs)
  fires involving, dioxin contamination by, 185–189
  structure of, 46
Polychlorinated dibenzofurans (PCDFs)
  background levels of, 264
  effects of, 265
  formation of, from chlorobenzene pyrolysis, 177
  structure of, 46
Polychlorinated quaterphenyls (PCQs), structure of, 46
Polychlorodibenzofurans, formation of , from chlorobenzene pyrolysis, 177
Polychlorophenols
  neutral impurities in, 176
  phenolic impurities in, 176

Polyethylene, chlorobenzenes from incineration of, 179
Ponds, TCDD degradation, 108–119
Predioxin, formula of, 175
Pregnancy, halogenated dioxin effects on, 54–57
Prenatal development, genetic and environmental interaction in, 202–204

R

Reclamation
  of areas exposed to TCDD, 69–79
  environmental, examples of, 151–155
Rice bran oil, PCB contamination of, 45–49
Regulation, of halogenated dioxins, 283–284
Ruthenium tetroxide, oxidative control of chemical pollutants by, 139–147

S

Seveso (Italy)
  tetrachlorodibenzo-*p*-dioxin release at, 5-35, 44
    birth defects from, 218–221, 271–280
    cancer incidence, 221
    chloracne incidence, 225
    clinical studies, 221–222
    decontamination of, 286–287
    health impact, 215–225
    neurological studies, 221–222
    occupational medicine, 223–224
Soil
  TCDD levels in,
    persistence, 33
    at Seveso, 7–16, 17–20
Sunlight effectivity against TCDD, 155
Symptoms of exposure to halogenated dioxins, 49–54, 215–225

T

2,4,5-T exposure studies, 236–237
Taiwan
  exposure to halogenated dioxins in, 45–46
    clinical symptomatology of, 49–54
Teratological effects of halogenated dioxins, 201–214, 218–221, 282
Tetrachloro isomers, dissipation time of, 59
Tetrachlorodibenzo-*p*-dioxin (TCDD), *see also* Halogenated dioxins
  accidental release of, at Seveso (Italy), 5–35
  agriculture problems from, 233–240
  in animals at Seveso, 27–29
  atmospheric levels at Seveso, 16, 21–23
  birth defects from, 218–221, 271–280
  carcinogenesis mechanisms related to, 191–195, 221
  clinical studies on, 221, 222
  from controlled combustion, 262–263
  environmental effects of release at Seveso, 31–35
  excretion products of, 84–88, 96
  fertility and, 207
  in fish, human exposure to, 252–255
  health impact of, 215–225, 229–230
  in herbicides, 260
  human exposure to, 250–255, 265
  in industrial wastes, 261
  LD_{50} values for, 83
  mechanism of action of, 265
  metabolism of, in mammals, 81–100
    by hepatic microsomes, 88–90
    role in toxicity, 94–96
  metabolites of, chemical structure, 93–94
  microbial degradation of, 105–135
    in aquatic and terrestrial model systems, 119–127
    in outdoor pond and model ecosystem, 108–119
  neurological effects of, 223
  photochemical degradation of, 180
  in plant tissues at Seveso, 27–29
  reclamation of area contaminated with, 69–79
  soil levels of, persistence, 33
    at Seveso, 7–16, 17–20
  toxicological data on, 204–208
    animal-to-human predictability, 208–209
  water levels of, at Seveso, 16
Tissues, halogenated dioxin effect on, 57–59
Trichlorophenol (TCP) synthesis, accident at Seveso during, 31–32

U

USDA, studies on 2, 4-D exposure, 237–239
UV sources, in halogenated dioxin degradation, 155

V

Vietnam
  Agent Orange use in, 229–230
    effects, 235

W

Water, tetrachlordibenzo-*p*-dioxin in, at Seveso, 16

Wildlife animals, TCDD levels in at Seveso, 30–31

Wood, burning of, chlorinated chemicals from, 177, 179

Y

Yusho (Japan), exposure to halogenated dioxins at 45–64

ECOTOXICOLOGY AND ENVIRONMENTAL QUALITY SERIES

Series Editors: *Frederick Coulston*
and
Freidhelm Korte

OTHER VOLUMES IN THE SERIES

Water Quality: Proceedings of an International Forum
F. Coulston and E. Mrak, editors

Regulatory Aspects of Carcinogenesis and Food Additives: The Delaney Clause
F. Coulston, editor

Environmental Lead
Donald R. Lynam, Lillian G. Piantanida, and Jerome F. Cole, editors

Metabolic Maps of Pesticides
Hiroyasu Aizawa, editor

Accidental Exposure to Dioxins: Human Health Aspects
Frederick Coulston and Francesco Pocchiari, editors

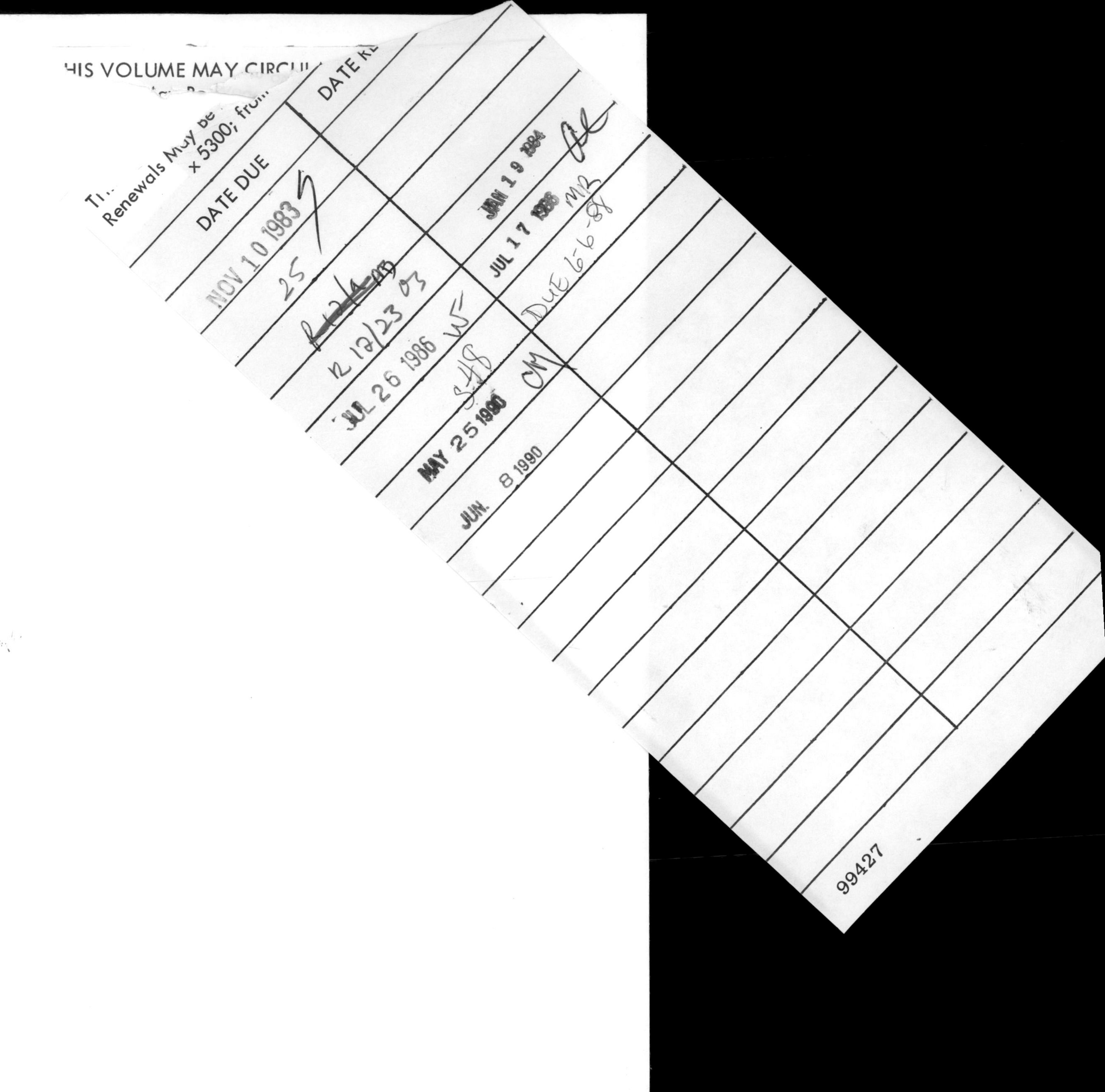

HIS VOLUME MAY CIRCUL
Renewals May be
x 5300; from
DATE DUE
DATE RE
NOV 10 1983
JAN 19 1984
JUL 17 1988
JUL 26 1986
MAY 25 1990
JUN. 8 1990
99427